Unsupervised Learning Approaches for Dimensionality Reduction and Data Visualization

Unsupervised Learning Approaches for Dimensionality Reduction and Data Visualization

B. K. Tripathy, Anveshrithaa Sundareswaran, and Shrusti Ghela

CRC Press
Taylor & Francis Group
Boca Raton London New York

CRC Press is an imprint of the
Taylor & Francis Group, an **informa** business

First edition published 2022
by CRC Press
6000 Broken Sound Parkway NW, Suite 300, Boca Raton, FL 33487-2742
and by CRC Press

2 Park Square, Milton Park, Abingdon, Oxon, OX14 4RN

Library of Congress Cataloging-in-Publication Data
A catalog record for this title has been requested

ISBN: 978-1-032-04101-8 (hbk)
ISBN: 978-1-032-04103-2 (pbk)
ISBN: 978-1-003-19055-4 (ebk)

Typeset in Times LT Std
by KnowledgeWorks Global Ltd.

The book is dedicated at the lotus feet of Lord Sri jagannath
for all his blessings bestowed upon me through out my life.

B. K. Tripathy

I would like to dedicate this book to my beloved parents,
Sundareswaran and Suganthy, whose unconditional love and support
have made me who I am today, and to the brave soldiers of the Indian
Army for their sacrifices and service to my motherland and its people.

Anveshrithaa Sundareswaran

For my parents, Shilpa Ghela and Bharat Ghela, who raised
me to believe anything was possible, and for my brother,
Hemin Ghela, whose love for technology and engineering
equals, and knowledge of it exceeds, my own.

Shrusti Ghela

Contents

Preface

We now live in a world where there is a mind-boggling amount of data available, a world that is becoming increasingly data-centric. With the dominance of data in every field, we strive to derive more value from this valuable resource. With the ever-growing amount of data on the one hand and the ever-advancing field of machine learning on the other, we are able to make far-reaching developments in every domain, ranging from science, healthcare, retail and finance to even art. Today, businesses rely heavily on their data with the growing need to analyze huge volumes of data to gain meaningful and actionable insights to achieve a competitive advantage and increase their business value. In the wake of this situation, data visualization and exploratory data analysis have become very important. The intricate task of extracting information from real world datasets that are generally large in volume with high dimensionality and discovering compact representations of such high dimensional data is the motivation behind the formulation of dimensionality reduction techniques. The principal idea of dimensionality reduction is to map data from a high dimensional space to a low dimensional space for the purpose of data visualization.

Taking this as the motivation, we present to you this book that provides a comprehensive discussion on how unsupervised learning can be exploited to solve the problem of dimensionality. Considering the importance of dimensionality reduction and data visualization in today's data-centric world, this book discusses various different unsupervised learning approaches to dimensionality reduction and data visualization. While classical dimensionality reduction techniques like PCA are limited to linear data, algorithms like Kernel PCA, Locally Linear Embedding, Laplacian Eigenmaps, Isomaps, Semidefinite Embedding, and t-SNE have been proposed that work well on nonlinear data. This book is a compilation of such algorithms. The principal objective of this book is to provide the readers a clear and intuitive understanding of these algorithms. This book is structured in a way that each algorithm is divided into separate chapters elaborately discussing the intuition behind the algorithms and their working. Lucid explanation of the mathematical concepts behind these algorithms along with derivations and proofs are also included to leave the reader with a solid theoretical as well as intuitive understanding of the algorithms. Each algorithm is explained with its original paper as the base, and all the important and necessary information presented in the papers is provided, but in a much more comprehensible way with simplified mathematical explanations for the reader's easy understanding. Each chapter is carefully curated by referring to various research papers and high-level technical notes on the subject to provide a better insight and in-depth understanding of the topic. This book reviews the strengths and limitations of each algorithm and highlights some of the algorithm's important use cases. Illustrative examples using datasets and visualizations of the data after dimensionality reduction are provided for every algorithm to give the reader an overview of how each algorithm works on the given datasets. With each example, a hands-on tutorial with step-by-step explanation and code for implementation is provided. Finally, this book includes a comparative study of all the algorithms discussed to give an idea of which algorithm will be the best choice for dimensionality reduction for a given dataset. This will help the reader to choose the most suitable and efficient algorithm for a given real-world problem. A glossary of words and concepts has been added to expand upon intricate and unfamiliar terms that are used in this book.

Overall, this book will give an extensive understanding of each of the algorithms by providing a holistic view of the algorithms in an easily comprehensible manner. This book is primarily intended for juniors, undergraduate and graduate level students majoring in computer science or related fields, and also for industry professionals and practitioners with a computer science or related background; it will give a good start to those who are looking forward to learning dimensionality reduction techniques for data visualization, either for the purpose of gaining theoretical knowledge on this subject or for practical application to work on projects. While this book provides a simplified view of various dimensionality reduction techniques, the reader is expected to have a basic knowledge of statistics, linear algebra and machine learning as a prerequisite for a better understanding of the subject.

B. K. Tripathy
Anveshrithaa Sundareswaran
Shrusti Ghela

Authors

Dr. B. K. Tripathy, a distinguished Researcher in mathematics and computer science, has more than 600 publications to his credit in international journals, conference proceedings, chapters in edited research volumes, edited volumes, monographs and books. He has supervised more than 50 research degrees to his credit. He has a distinguished professional career of more than 40 years of service in different positions, and at present he is working as a Professor (higher academic grade) and as Dean of the school of Information Technology at VIT, Vellore. As a student, Dr. Tripathy won three gold medals, a national scholarship for post graduate studies, a UGC fellowship to pursue his PhD, DST sponsorship to pursue his M. Tech in computer science at Pune University, and a DOE visiting fellowship to IIT, Kharagpur. He was nominated as a distinguished alumni of Berhampur University on its silver jubilee and golden jubilee years. For efficient service as a reviewer for *Mathematical Reviews*, he was selected as an honorary member of the American Mathematical Society. Besides this he is a life member/senior member/member of more than 20 international professional societies including IEEE, ACM, IRSS, CSI, Indian Science Congress, IMS, IET, ACM Compute News group and IEANG. Dr. Tripathy is an editor/editorial board member/reviewer of more than 100 international journals, including *Information Sciences, IEEE Transactions on Fuzzy Systems, Knowledge Based Systems, Applied Soft Computing, IEEE Access, Analysis of Neural Networks, International Journal of Information Technology and Decision Making, Proceedings of the Royal Society-A* and *Kybernetes*. He has so far adjudicated PhD theses of more than 20 universities all over India. He has organized many international conferences, workshops, FDPs, guest lectures, industrial visits and webinars over the past 14 years. Dr. Tripathy has to his credit delivered keynote speeches in international conferences, organized special sessions and chaired sessions. Also, many of his papers have been selected as best papers at international conferences. He has received funded projects from UGC, DST and DRDO and published some patents also.

The current topics of research of Dr. Tripathy include soft computing, granular computing, fuzzy sets and systems, rough sets and knowledge representation, data clustering, social network analysis, neighbourhood systems, soft sets and decision making, social internet of things, big data analytics, multiset theory, decision support systems, deep neural networks, Healthcare analytics, pattern recognition and dimensionality reduction.

Anveshrithaa Sundareswaran completed her undergraduate degree in Computer Science and Engineering at Vellore Institute of Technology, Vellore. She is enrolled as a graduate student pursuing master's degree in Computational Data Science from the School of Computer Science at Carnegie Mellon University. Her areas of interest include machine learning, deep learning and data science. She has shown her research capabilities with several publications. Her research on "Promoter Prediction in DNA Sequences of Escherichia coli using Machine Learning Algorithms" won the

Best Student Paper award at the IEEE Madras section, Student Paper Contest, 2019 and was published later in the *International Journal of Scientific and Technology Research*. She presented a paper on "Real-Time Vehicle Traffic Analysis using Long Short-Term Memory Networks in Apache Spark" at the IEEE International Conference on Emerging Trends in Information Technology and Engineering, 2020. Her research on "Real-Time Traffic Prediction using Ensemble Learning for Deep Neural Networks" has been published in the *International Journal of Intelligent Information Technologies (IJIIT)*. Also, she has communicated a research paper on "Real-Time Weather Analytics using Long Short-Term Memory Networks" to the *International Journal of Cognitive Computing in Engineering.* Her other achievements include the Outstanding Student award at the 2020 Tsinghua University Deep Learning Summer School where she was the only student to represent India. The Achievers Award and Raman Research Award from VIT University are some of the other recognitions of her merit.

Shruthi Ghela received her B. Tech (CS) degree from Vellore Institute of Technology, Vellore in May 2020. She completed her Capstone project during her time at Iconflux Technologies Pvt. Ltd., Ahmedabad, India in the field of machine learning. She has completed two summer projects: one in the domain of data science for KeenExpert Solution Pvt. Ltd., Ahmedabad, India, and the other in web development for Jahannum.com, Ahmedabad, India. For her excellent academic performance, she received a scholarship for all four years of under graduation from VIT. She was the winner of the DevJams'19 hackathon for two consecutive years in 2018 and 2019. She has proficiency in the German and Chinese languages, as well as English. In an attempt to increase and intensify her specializations, she has completed the IBM Data Science Professional Certificate (Coursera), Machine Learning A–Z (Udemy), and Machine Learning by Andrew Ng (Coursera). Ms. Ghela has the skill set of Hadoop, Python, MATLAB®, R, Haskell, object oriented programming, full stack development, functional programming and statistics. Her research area of interest includes data science and quantum computing. Along with being a hard-core subject learner, she enjoys photography, playing tennis, reading and traveling.

1 Introduction to Dimensionality Reduction

1.1 INTRODUCTION

The world is currently witnessing unprecedented technological advancements that are changing the world in every way possible. We are already in the era of data, where the amount of data that is available and the rate at which data is being generated are outside of the realm of our imagination. The amount of data that is being created every second is growing at an exponential rate resulting in data explosion and this growth is expected to be ever increasing. Saying so, with exponentially more data than ever, we are moving toward a data-centric world, where every field is becoming data dominated. Data being one of the most valuable resources today, we are working toward deriving more value from it, as a wealth of information can be processed from this data that can have a profound impact on shaping the world in many aspects.

On the other hand, the evolution of machine learning is happening at a tremendous pace. The advent of machine learning and AI has brought in one of the greatest technological transformations that is revolutionizing the world. The idea that machines can learn from data with minimal human interference has become so prevalent that machine learning has a ubiquitous influence in almost every field. Furthermore, with the inception of high power computing leading to the increased availability of cheaper and more powerful computational capacity and affordable data storage, it is possible to make extensive progress in various fields, by leveraging machine learning to make good use of the available data. This is also leading to major breakthroughs in the field of science and research. For instance, understanding biological data by exploiting machine learning is greatly helping in finding answers to some of the important questions on the existence of life on earth, causes of diseases, and the effect of microorganisms on the human body; this is also leading to remarkable findings in cancer research and drug discovery. Furthermore, analysis of space data has also paved the way for noteworthy and fascinating findings in the field of space exploration and is greatly helping in solving some of the mysteries that are still beyond human imagination. Coming to the finance industry, analysis of large amounts of data plays a vital role as it is crucial to identify important insights in data to learn about trends in stock prices, investment and trade opportunities, identifying high-risk profiles, and preventing fraud. Another sector that is extensively using massive amounts of data to derive value from them is the healthcare industry. Large volumes of real-time data from sensors and wearable devices are analyzed to identify trends, diagnose diseases, and even treat them. This list is endless as data science plays an important role in almost every domain ranging from computational genomics to environmental science to even art.

Today, businesses heavily rely on their data for their sustainability. Data has become an inevitable aspect of every business today with the growing demand to analyze the data to derive important insights for better decision making and improving business strategies. The adoption of machine learning by enterprises for analysis of their data to answer questions from that data has accelerated over the last few years. Enterprises are capitalizing on data to accelerate their businesses and increase their efficiency, productivity, and profit. Companies are largely investing in data to unleash the potential of their data in improving their business value.

Owing to this situation, data visualization and exploratory data analysis have gained tremendous importance. But extracting information from real-world datasets that are generally large in volume with high dimensionality and discovering compact representations of such high dimensional data is an intricate task. Sometimes, the data can be of extremely large dimensions, on the order of thousands or sometimes even millions which makes it infeasible for analysis as many algorithms perform poorly on very high dimensional data. Given that the volume of data itself is very large, the problem of high dimensionality is not trivial, making it necessary to find an approximation to the original data that has much fewer dimensions but at the same time retains the structure and the relationships within the data.

The main goal of unsupervised learning is to create compact representations of the data by detecting their intrinsic and hidden structure for applications in data visualization and exploratory data analysis. The need to analyze large volumes of multivariate data for tasks like pattern recognition, analysis of gene expression data and time series analysis across a wide range of fields raises the fundamental issue of discovering compact representations of high dimensional data. This difficulty in extracting information from high dimensional data is the principal motivation behind the renewed interest in formulating the problem of dimensionality reduction. The idea behind dimensionality reduction is to map data in a higher dimensional space to a lower dimensional space as a method for data visualization and to extract the key low dimensional features. While classical unsupervised dimensionality reduction approaches like Principal Component Analysis (PCA) are limited to linear data in terms of effectiveness as they overlook correlations in the data that are higher than second order, relatively new approaches like Locally Linear Embedding (LLE), Laplacian Eigenmaps, Isomap, Semidefinite Embedding, and t-distributed stochastic neighbor embedding (t-SNE) have been proposed in the recent past as solutions to resolving the problem of dimensionality reduction in the case of nonlinear relationships within the data.

Linear methods such as PCA and Multidimensional Scaling (MDS) are some of the fundamental spectral methods for linear dimensionality reduction whose underlying geometric intuitions form the basis for many other nonlinear dimensionality reduction techniques.

In Chapter 1, we discuss one of the oldest and most widely known dimensionality reduction techniques, PCA, which is based on linear projection. This is a fundamental, classical approach to dimensionality reduction, but its major limitation is its inability to handle nonlinear data. This technique projects data to lower dimensional space by maximizing the variance. The principal components are obtained by finding the solution to an eigenproblem, that is, by performing singular value decomposition on the data matrix. An algorithm which is a slight variation of the classical PCA, called the Dual PCA, is discussed in Chapter 2.

Since PCA is restricted to linear dimensionality reduction where high dimensional data lies on a linear subspace, a kernel-based method which is a nonlinear generalization of PCA was introduced to handle nonlinear data. This method is based on the "kernel trick" where the inner products are replaced with a kernel. The kernel function can be considered as a nonlinear similarity measure and many linear approaches can be generalized to nonlinear methods by exploiting the "kernel trick." This variant of PCA uses kernels to compute the principal components. It works well on nonlinear data and overcomes the limitation of PCA in addressing the nonlinearity of the data. This technique came to be known as the kernel PCA, which is extensively discussed in Chapter 3.

While PCA is an efficient dimensionality reduction technique, in cases where the dimensionality of the data is very high, PCA becomes computationally expensive. Chapter 4 discusses another dimensionality reduction technique called random projection that uses a projection matrix whose entries are randomly sampled from a distribution to project the data to a low dimensional space and also guarantees pairwise distance preservation in the low dimensional space. This method of dimensionality reduction outperforms PCA when the dimensionality of the data is very high which makes PCA computationally expensive.

In Chapter 5, we will discuss Canonical Correlation Analysis which is used as a dimensionality reduction technique for multi-view setting, that is, when there are two or more views of the same data. The goal is to generate low dimensional representations of the points in each view such that it retains the redundant information between the different views by singular value decomposition.

Similar to PCA, MDS is yet another classical approach to dimensionality reduction that attempts to preserve the pairwise distances between the data points. Again, MDS is a linear dimensionality reduction technique just like PCA and is limited to linear data. This method of mapping data points from high dimensional space to low dimensional space is elaborately explained in Chapter 6.

These linear methods, such as PCA and MDS, produce good low dimensional representations when the geometry of the original input data is confined to a low dimensional subspace, whereas, in the case where the input data is sampled from a low dimensional submanifold of the input space which makes it nonlinear, these linear methods fail to perform well. Many powerful graph-based methods have been proposed for nonlinear dimensionality reduction. These manifold learning approaches construct graphs representing the data points and the relationships between them from which matrices are formed whose spectral decomposition gives low dimensional representations of the nonlinear data. Some of these graph-based manifold learning approaches to nonlinear dimensionality reduction are discussed in the subsequent chapters.

Just like PCA, MDS has also been extended to solve the problem of nonlinear dimensionality reduction. This approach, one of the earliest manifold learning techniques, attempts to unfold the intrinsically low dimensional manifold on which the data points lie and uses the geodesic distances between data points rather than Euclidean distance to find the low dimensional mapping of the points such that the geodesic distances between pairs of data points are preserved. This nonlinear generalization of MDS, called as Isomap is discussed in Chapter 7.

The method of LLE proposed in 2000 by Sam T. Roweis and Lawrence K. Saul is a nonlinear dimensionality reduction technique that identifies the nonlinear structure in the data and finds neighborhood preserving mapping in low dimensional space. A detailed discussion on how this algorithm works and how the local distances are preserved is presented in Chapter 8.

Clustering techniques are used to identify groups within data that exhibit some similarity within them and one such algorithm is the spectral clustering method which is discussed in detail in Chapter 9. Though it is not a dimensionality reduction technique by itself, it has close connection with another technique called the Laplacian eigenmap which is a computationally efficient approach to nonlinear dimensionality reduction that uses the concept of graph Laplacians to find locality preserving embeddings in the feature space. This algorithm and its association with spectral clustering are elaborated upon in Chapter 10.

While kernel PCA is the nonlinear generalization of PCA that uses kernel function in place of inner product, the choice of the kernel is very important in that it determines the low dimensional mappings. Unlike the kernel PCA where the kernel that is going to be used is determined prior, Maximum Variance Unfolding, also known as Semidefinite Embedding, is an algorithm that learns the optimal kernel by semidefinite programming. This form of kernel PCA that is different from other manifold learning methods like LLE and Isomap is elaborately dealt with in Chapter 11.

A relatively new method for visualizing high dimensional data is the t-SNE proposed by Laurens van der Maaten and Geoffrey Hinton in the year 2008. This algorithm is a variation of an already existing algorithm called stochastic neighbor embedding (SNE) and was proposed in an attempt to overcome the limitations of SNE. Comparatively, t-SNE is much easier to optimize and yields significantly better embeddings than SNE. While many algorithms attempt to preserve the local geometry of the data, most of them are not capable of capturing both the local and global structure. But t-SNE was able to resolve this issue and provide better visualizations than other techniques. An extensive discussion on how t-SNE works is presented in Chapter 12.

To put it in a nutshell, in each chapter, we will elaborately discuss these unsupervised learning approaches to the problem of dimensionality reduction by understanding the intuition behind these algorithms using theoretical and mathematical explanations. We will also evaluate each of these algorithms for their strengths and weaknesses. The practical use cases of the algorithms are briefly discussed. Additionally, illustrative examples with visualizations of data are provided for each technique along with a hands-on tutorial that includes step-by-step explanation with code for dimensionality reduction and visualization of data. Finally, a section exclusively for a comparative analysis of all these algorithms is presented to help in assessing the performance of each of the algorithms on a given dataset and choosing the best suitable algorithm for the purpose of dimensionality reduction of the data.

2 Principal Component Analysis (PCA)

2.1 EXPLANATION AND WORKING

Principal Component Analysis (PCA), a feature extraction method for dimensionality reduction, is one of the most popular dimensionality reduction techniques. We want to reduce the number of features of the dataset (dimensionality of the dataset) and preserve the maximum possible information from the original dataset at the same time. PCA solves this problem by combining the input variables to represent it with fewer orthogonal (uncorrelated) variables that capture most of its variability [1].

Let the dataset contain a set of n data points denoted by $x_1, x_2, ..., x_n$ where each x_i is a d-dimensional vector. PCA finds a p-dimensional linear subspace (where $p < d$, and often $p \ll d$) in a way that the original data points lie mainly on this p-dimensional linear subspace. In practice, we do not usually find a reduced subspace where all the points lie precisely in that subspace. Instead, we try to find the approximate subspace which retains most of the variability of data. Thus, PCA tries to find the linear subspace in which the data approximately lies.

The linear subspace can be defined by p orthogonal vectors, say: $U_1, U_2, ..., U_p$. This linear subspace forms a new coordinate system and the orthogonal vectors that define this linear subspace are called the "principal components" [2]. The principal components are a linear transformation of the original features, so there can be no more than d of them. Also, the principal components are perpendicular to each other. However, the hope is that only p $(p < d)$ principal components are needed to approximate the d-dimensional original space. In the case, where $p = d$ the number of dimensions remains the same and there is no reduction.

Let there be n data points denoted by $x_1, x_2, ..., x_n$ where each x_i is a d-dimensional vector. The goal is to reduce these points and find a mapping $y_1, y_2, ..., y_n$ where each y_i is a p-dimensional vector (where $p < d$, and often $p \ll d$). That is, the data points $x_1, x_2, ..., x_n \ \forall \ x_i \in R^d$ are mapped to $y_1, y_2, ..., y_n \ \forall y_i \in R^p$.

Let X be a $d \times n$ matrix that contains all the data points in the original space which has to be mapped to another $p \times n$ matrix Y (matrix), which retains maximum variability of the data points by reducing the number of features to represent the data point.

Note: In PCA or any variant of PCA, a standardized input matrix is used. So, X represents the standardized input data matrix, unless otherwise specified.

Now let us discuss how PCA solves the problem of dimensionality reduction. PCA is a method based on linear projection. For any linear projection-based techniques, given a d-dimensional vector x_i, we obtain a low dimensional representation y_i (a p-dimensional vector) such that

$$y_i = U^T x_i \tag{2.1}$$

How is the direction of the principal components chosen? The basic idea is to pick a direction along which data is maximally spread, that is, the direction along which there is maximum variance [1, 3, 4].

Let us consider the first principal component (the direction along which the data has maximum variance) to be U_1. We now project the n points from matrix X on U_1, given by:

$$U_1^T X \tag{2.2}$$

So, by definition, we would now like to find the maximum variance of $U_1^T X$. To find that we solve for the below optimization problem:

$$\max_{U_1} U_1^T X$$

We know that

$$var(U_1^T X) = U_1^T \Sigma U_1 \tag{2.3}$$

Here, Σ is a sample covariance matrix of the data matrix X. Thus, the optimization problem now becomes

$$\max_{U_1} U_1^T \Sigma U_1 \tag{2.4}$$

However, $U_1^T \Sigma U_1$ is a quadratic function with no fixed upper bound. So, (2.4) turns out to be an ill-defined problem. This is because (2.4) depends on the direction as well as the magnitude of U_1. To convert this into a well-defined problem, we need to add a constraint to (2.4).

There are two possible approaches to resolve this problem. Either we add a constraint on the direction of U_1 or on the magnitude of U_1. Adding a constraint on the direction of U_1 and trying to calculate $max\,(U_1^T \Sigma U_1)$ will still make it an ill-defined problem, because there is no upper-bound even after adding the constraint. But if we add a constraint on the magnitude of U_1, that is, if we restrict the magnitude of U_1, let us say $U_1^T U_1 = 1$, the length of the principal component is fixed. Hence, there is only one direction in which $U_1^T \Sigma U_1$ would be maximum. Thus, this problem has an upper bound, and to solve for the upper bound, we are interested in finding the direction of U_1.

Using the second case, between all fixed-length vectors, we search for the direction of the maximum variance of the data. Now we have a well-defined problem as follows:

$$\max_{U_1} U_1^T \Sigma U_1$$
$$\text{subject to } U_1^T U_1 = 1 \tag{2.5}$$

Generally, when we want to maximize (or minimize) a function subject to a constraint, we use the concept of Lagrange multipliers.

Lagrange's multipliers say that there exists $\lambda_1 \in \boldsymbol{R}$ such that the solution U_1 to the above problem can be rewritten as:

$$L(U_1, \lambda_1) = U_1^T \Sigma U_1 - \lambda_1 (U_1^T U_1 - 1) \qquad (2.6)$$

To optimize (2.6), we simply take the derivative and equate it to 0. This gives:

$$\Sigma U_1 - \lambda_1 U_1 = 0 \qquad (2.7)$$

Thus,

$$\Sigma U_1 = \lambda_1 U_1 \qquad (2.8)$$

Here, Σ is the covariance matrix of X, λ_1 is the dual variable and the vector that we are looking for is U_1.

The direction U_1 obtained by maximizing the variance is the direction of some unit vector that satisfies (2.8). However, this is exactly the definition of an eigenvector of matrix Σ. Note that the eigenvector of the matrix multiplied to the matrix results in a vector that is just the scaled version of the eigenvector itself. So, U_1 is the eigenvector of Σ, and λ_1 is the corresponding eigenvalue.

Now, the question is which eigenvector to choose? Σ is a $d \times d$ matrix. So, there are at most d eigenvectors and eigenvalues.

Now, our objective is to maximize $U_1^T \Sigma U_1$ and from (2.8) we know that $\Sigma U_1 = \lambda_1 U_1$:

$$U_1^T \Sigma U_1 = U_1^T \lambda_1 U_1 \qquad (2.9)$$

$$= \lambda_1 U_1^T U_1 \qquad (2.10)$$

$$= \lambda_1 \qquad (2.11)$$

And since we want to maximize the above quantity, that is, we want to maximize λ_1, it is evident that we have to pick U_1 to be the eigenvector corresponding to the largest eigenvalue of Σ. For U_2, ..., U_p we proceed similarly and pick the eigenvector that has the second largest, up to the pth largest eigenvalues. Thus, the eigenvector of the sample covariance matrix Σ corresponding to the maximum eigenvalue is the first principal component. Similarly, the second principal component is the eigenvector with the second largest eigenvalue and so on. In the same way, all the principal components can be found just by the eigendecomposition of the covariance matrix [5].

Singular Value Decomposition (SVD) could quickly solve the eigendecomposition problem in a computationally efficient manner. SVD routines are much more numerically stable than eigenvector routines for finding eigenvectors in the matrix. We can use SVD to calculate eigenvectors in PCA very efficiently [5, 6]. SVD is a matrix factorization technique which expresses a matrix (in this case, it is X) as a linear combination of matrices of rank 1. SVD is a stable method because it does not require a positive definite matrix [6].

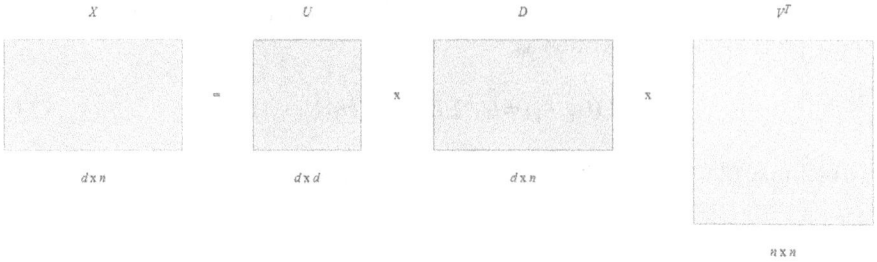

FIGURE 2.1   Representation of singular value decomposition when $d > n$.

Figure 2.1 represents the case when the number of data points is greater than the dimensionality of any data point $(n > d)$. As shown in Figure 2.1, SVD produces three matrices U, D, and V^T. U and V^T are orthogonal matrices whose columns represent orthonormal eigenvectors of XX^T and X^TX. The matrix D is a diagonal matrix (diagonal values are known as singular values). Each singular value is the square root of the corresponding eigenvalues of U or V^T (both matrices have the same positive eigenvalues) [5]. Matrix U represents the eigenvectors of XX^T, which is the covariance matrix Σ. Hence, we could easily find the top p eigenvectors of the covariance matrix Σ (which is the solution for PCA) using Singular Value Decomposition on the input matrix X [6].

2.2   ADVANTAGES AND LIMITATIONS

- Correlated features are removed: When we are working with data in the real world, it is prevalent that we have thousands of features in the dataset. So, if we run any algorithm on the dataset using all the features, it will reduce the algorithm's performance, and it will not be easy to visualize that many features in any kind of graph. So, we need to reduce the number of features in the dataset. One way to do this is to determine the correlation between the features (correlated variables). To manually find correlation between thousands of features is nearly impossible, frustrating, and time-consuming [6]. PCA solves this problem efficiently. After implementing PCA on the dataset, all the principal components are independent of one another, and there is no correlation between them.
- Algorithm performance is improved: With so many features, the performance of the algorithm will drastically degrade. PCA is a technique that can help in speeding up machine learning algorithms by getting rid of correlated variables that do not contribute to any decision making. With a smaller number of features, the training time of the algorithm decreases significantly. So, if the input dataset dimension is enormous, then PCA helps us speed up our algorithm.
- Overfitting is reduced: Overfitting usually occurs when there are too many variables in the dataset. So, PCA helps in overcoming the overfitting issue by reducing the number of features [6].

- Visualization is improved: It is tough to visualize and understand the data in high dimensions [6]. PCA helps us transform high dimensional data to low dimensional data to visualize it easily.
- While PCA is a widely used technique for dimensionality reduction, it has some drawbacks as well. Firstly, the model performance of PCA drastically reduces if the assumptions of linearity are not met. PCA is limited to linear data in terms of effectiveness and fails when there are nonlinear relationships within the data.
- Furthermore, the independent variables become less interpretable. After we implement PCA on the dataset, the original features turn into principal components. Principal components are just linear combinations of the features of the original dataset [3]. Thus, principal components are not as readable and understandable as the original features.
- Moreover, data standardization is a must before PCA. Before implementing PCA, We have to standardize the data. Otherwise, PCA will not find the correct principal components. For example, let us say we have a feature set with data denoted in units of Kilograms, Light years, or Millions [6]. In such a dataset, the variance scale is enormous. If PCA is applied to this kind of dataset, the resultant loadings for high variance features will also be large. Thus, principal components will be biased toward high variance features, which leads to false results. Also, for standardization, all the categorical variables have to be converted into numerical values before applying PCA.
- Even though principal components try to cover maximum variance among the variables in the dataset, if we do not select the number of principal components with care, there is a good chance that we might miss some information as compared to the original list of variables [6].
- Model performance reduces if the datasets have no or very low feature correlation.
- The variance-based PCA technique does not consider the differentiating characteristics of the different classes. Furthermore, if the information that differentiates one class from another is in the low variance components, it might be discarded.

Extensions could solve some of these specific limitations to classical PCA. Some of these extensions include Randomized PCA, Sparse PCA, Dual PCA, Kernel PCA, and many more. Randomized PCA quickly approximates the first few principal components in very high dimensional data. Sparse PCA introduces a regularization tuning parameter for increased sparsity. We will discuss Dual PCA and Kernel PCA in detail in Chapter 3 and Chapter 4, respectively.

2.3 USE CASES

Essentially PCA is used as a dimensionality reduction technique in various domains such as facial recognition, computer vision, and image compression. Furthermore, it is also used to find patterns in data of high dimension in finance, data mining, bioinformatics, psychology, and many more [5]. We discuss a few applications of PCA briefly below:

- Neuroscience:
    - A technique, commonly known as spike-triggered covariance analysis, uses a variant of PCA in neuroscience to identify some particular properties of a stimulus that increases a neuron's probability of generating an action potential.
    - It is also used to find the identity of a neuron from the shape of its action potential.
    - PCA is used to reduce dimensionality to detect coordinated activities of large neuronal ensembles. It is used to determine collective variables and order parameters during phase transitions in the brain.

- Quantitative Finance:
    - PCA is used to reduce dimensionality in quantitative finance. For example, let us say a fund manager has 200 stocks in his portfolio. To analyze these stocks quantitatively, a stock manager will require a co-relational matrix of 200×200, which makes the problem very complicated. However, if the fund manager extracts 10 principal components, which best represent the variance in the stocks, this would reduce the complexity of the problem while still explaining all 200 stocks' movements.
    - Moreover, PCA is used to analyze the shape of the yield curve, hedging fixed income portfolios, to implement interest rate models, to forecast portfolio returns, to develop asset allocation algorithms, to develop long-short equity trading algorithms.

- Facial Recognition:
    - EigenFaces (a classical method in computer vision) is used for facial recognition. According to Sirovich and Kirby, PCA could be used on a collection of face images to form a set of necessary features.
    - PCA forms the basis of the EigenFaces approach as the set of EigenFaces is generated using PCA.
    - The EigenFaces approach reduces statistical complexity in face image representation.
    - Other researchers have increased face recognition accuracy by using a combination of Wavelet, PCA, and neural networks.

- Other Applications:
    - PCA is used on medical data to show the correlation of cholesterol with low-density lipoprotein.
    - PCA is used on HVSR (horizontal to vertical spectral ratio) data aimed at earthquake-prone areas' seismic characterization.
    - PCA is used for the detection and visualization of computer network attacks.

- PCA is used for anomaly detection.
- PCA is also used for image compression.

2.4   EXAMPLES AND TUTORIAL

To better understand PCA's working, let us consider a few simple examples where we perform dimensionality reduction on a very high dimensional dataset and visualize the results.

Example 1

While most of the real-world datasets like images and text data are very high dimensional, we will use the MNIST handwritten digits dataset for simplicity. The MNIST dataset is a collection of grayscale images of handwritten single digits between 0 and 9 that contains 60,000 images of size 28 × 28 pixels. Thus, this dataset has 60,000 data samples with a dimensionality of 784. To demonstrate dimensionality reduction on this dataset, we use PCA to reduce the data's dimensionality and project the data onto a low dimensional feature space. This example will map the data with 784 features to two-dimensional feature space and visualize the results.

Let us import the MNIST handwritten digits dataset from the tensorflow library. Next, we will use the *sklearn.decomposition* module from the scikit-learn library for dimensionality reduction. Finally, after applying PCA on the dataset, we will plot the results to visualize the low dimensional representation of the data using the Matplotlib library.

The first step is to import all the necessary libraries.

```
import tensorflow as tf
from tensorflow.examples.tutorials.mnist import input_data
import sklearn
from sklearn.decomposition import PCA
import matplotlib.pyplot as plt
```

Next, let us load the MNIST dataset.

```
mnist = input_data.read_data_sets("MNIST_data/")
```

This dataset is split into three parts:

```
1. mnist.train - which has 55000 data points
2. mnist.test - which has 10000 data points
3. mnist.validation - which has 5000 data points
```

Every MNIST data point has two parts:

1. The handwritten digit image (X)
2. The corresponding class label (Y)

mnist.train.images contains all the training images and mnist.train.labels has all the corresponding training labels. As told earlier, each image is of size 28 × 28 pixels, which is flattened into a vector of size 784. Hence, mnist.train.images is an *n*-dimensional array (tensor) whose shape is [55000, 784], whereas, the shape of mnist.train.labels is [55000, 10] since there are 10 class labels from 0 to 9.

```
X_train = mnist.train.images
Y_train = mnist.train.labels
n_samples, n_features = X_train.shape
print(n_features)
print(n_samples)
```

Output:

```
784
55000
```

We use PCA. For this we use *sklearn.decomposition.PCA*. The parameter *n_components* denotes the dimensionality of the target projection space. We then fit and transform the training data.

```
pca =PCA(n_components=2)
pca.fit(X_train)
X_pca = pca.fit_transform(X_train)
```

Hence, the dimensionality of the projection space is reduced from 784 to 2. That is, the shape of X_pca is [55000, 2]

```
n_samples, n_features = X_pca.shape
print(n_features)
print(n_samples)
```

Output:

```
2
55000
```

Now, let us plot the reduced data using matplotlib where each data point is represented using a different color corresponding to its label.

```
plt.figure(figsize=(10,10))
plt.scatter(X_pca[y==0, 0], X_pca[y==0, 1], color='blue',
alpha=0.5,label='0', s=9, lw=2)
plt.scatter(X_pca[y==1, 0], X_pca[y==1, 1], color='purple',
alpha=0.5,label='1',s=9, lw=2)
plt.scatter(X_pca[y==2, 0], X_pca[y==2, 1], color='yellow',
alpha=0.5,label='2',s=9, lw=2)
plt.scatter(X_pca[y==3, 0], X_pca[y==3, 1], color='black',
alpha=0.5,label='3',s=9, lw=2)
plt.scatter(X_pca[y==4, 0], X_pca[y==4, 1], color='gray',
alpha=0.5,label='4',s=9, lw=2)
```

```
plt.scatter(X_pca[y==5, 0], X_pca[y==5, 1], color='turquoise',
alpha=0.5,label='5',s=9, lw=2)
plt.scatter(X_pca[y==6, 0], X_pca[y==6, 1], color='red',
alpha=0.5,label='6',s=9, lw=2)
plt.scatter(X_pca[y==7, 0], X_pca[y==7, 1], color='green',
alpha=0.5,label='7',s=9, lw=2)
plt.scatter(X_pca[y==8, 0], X_pca[y==8, 1], color='violet',
alpha=0.5,label='8',s=9, lw=2)
plt.scatter(X_pca[y==9, 0], X_pca[y==9, 1], color='orange',
alpha=0.5,label='9',s=9, lw=2)
plt.ylabel('Y coordinate')
plt.xlabel('X coordinate')
plt.legend()
plt.show()
```

Figure 2.2 shows the visualization of the MNIST data reduced to a two-dimensional feature space using PCA.

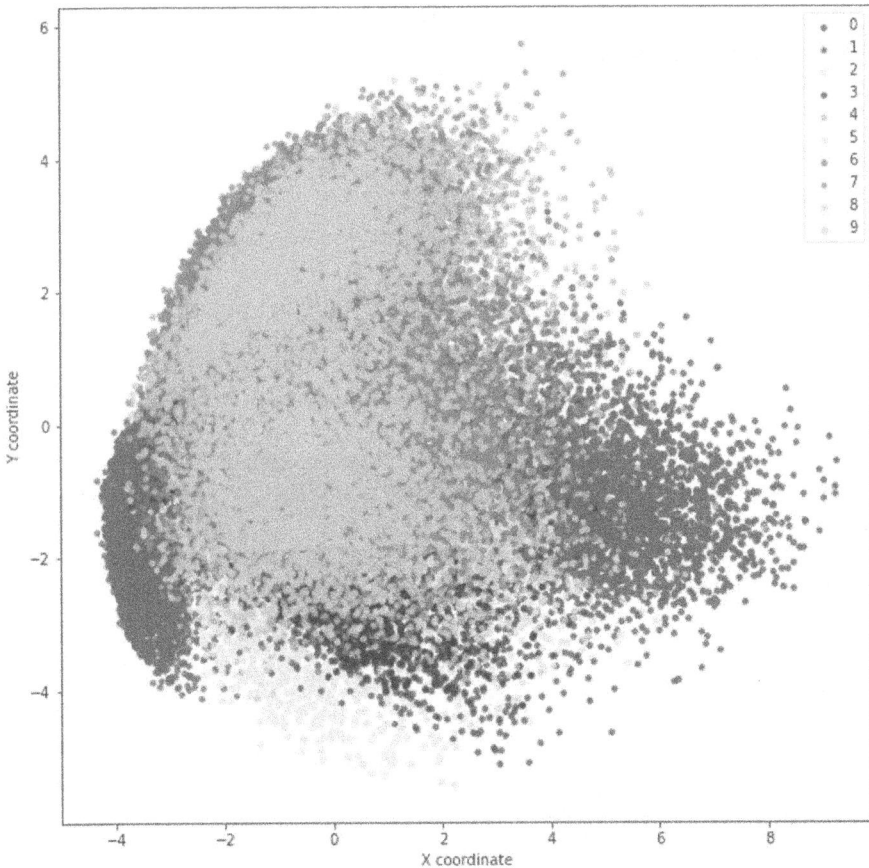

FIGURE 2.2   PCA on MNIST data.

Example 2

Next, let us see the results of dimensionality reduction on the 64-dimensional hand-written digits dataset (not to be confused with the MNIST handwritten digits database) consisting of 1083 images of six handwritten digits (0–5) each of size 8 × 8 pixels.

Let us import the digits dataset from the sklearn library. Similar to the previous example, we will use the *sklearn.decomposition* module from the Scikit-learn library for dimensionality reduction and matplotlib for plotting the data.

First, import all the necessary libraries.

```
import sklearn
from sklearn import datasets
from sklearn.decomposition import PCA
import matplotlib.pyplot as plt
```

Next, load the digits dataset using *load_digits*. This dataset originally contains 1797 data samples. That is, it contains 1797 images of digits. Each data point is an image of a digit of size 8 × 8. Hence, the dimensionality of the data is 64. The dataset is divided into 10 classes (digits from 0 to 9). Thus, the shape of the data is [1797, 64]. However, in this example we will consider only 6 classes (digits with label 0–5) which contain 1083 samples.

```
digits = datasets.load_digits(n_class=6)
X = digits.data
y = digits.target
n_samples, n_features = X.shape
print(n_features)
print(n_samples)
```

Output:

```
64
1083
```

Here, digits.data contains the images and digits.target contains the corresponding labels similar to mnist.train.images and mnist.train.labels, respectively (as in the previous example).

We use *sklearn.decomposition.PCA* for PCA. The parameter *n_components* denotes the dimensionality of the target projection space. We then fit and transform the training data.

```
pca = PCA(n_components=2)
X_pca = pca.fit_transform(X)
```

Now, plot the transformed data using matplotlib. Figure 2.3 visualizes the two-dimensional embedding obtained after performing dimensionality reduction on the 64-dimensional data consisting of 1083 data samples using PCA.

```
plt.figure(figsize=(10,10))
plt.scatter(X_pca[y==0, 0], X_pca[y==0, 1], color='blue',
alpha=0.5,label='0', s=9, lw=2)
```

```
plt.scatter(X_pca[y==1, 0], X_pca[y==1, 1], color='green',
alpha=0.5,label='1',s=9, lw=2)
plt.scatter(X_pca[y==2, 0], X_pca[y==2, 1], color='orange',
alpha=0.5,label='2',s=9, lw=2)
plt.scatter(X_pca[y==3, 0], X_pca[y==3, 1], color='purple',
alpha=0.5,label='3',s=9, lw=2)
plt.scatter(X_pca[y==4, 0], X_pca[y==4, 1], color='violet',
alpha=0.5,label='4',s=9, lw=2)
plt.scatter(X_pca[y==5, 0], X_pca[y==5, 1], color='red',
alpha=0.5,label='5',s=9, lw=2)
plt.ylabel('Y coordinate')
plt.xlabel('X coordinate')
plt.legend()
plt.show()
```

Figure 2.3 visualizes the two-dimensional embedding obtained after performing dimensionality reduction on the 64-dimensional data consisting of 1083 data samples using PCA.

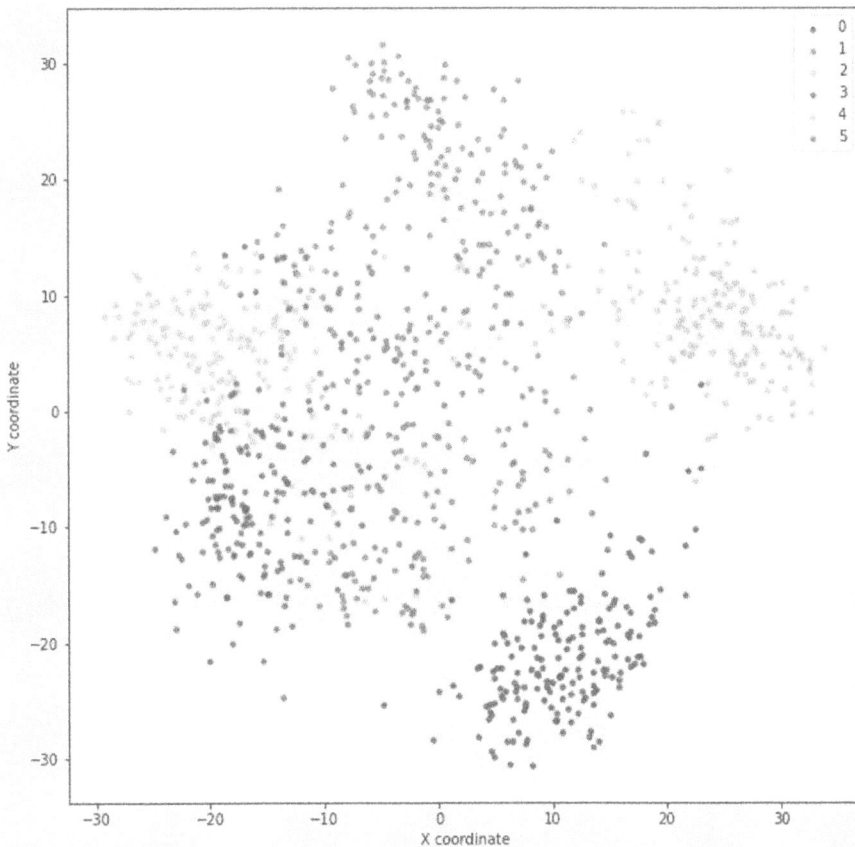

FIGURE 2.3 PCA on digits dataset.

REFERENCES

1. Wold, S., Esbensen, K., & Geladi, P. (1987). Principal component analysis. *Chemometrics and intelligent laboratory systems*, 2(1–3), 37–52.
2. Abdi, H., & Williams, L. J. (2010). Principal component analysis. *Wiley interdisciplinary reviews: computational statistics*, 2(4), 433–459.
3. Saul, L. K., Weinberger, K. Q., Sha, F., Ham, J., & Lee, D. D. (2006). Spectral methods for dimensionality reduction. *Semi-supervised learning*, 3, 293–308.
4. Burges, C. J. (2009). Geometric methods for feature extraction and dimensional reduction—a guided tour. In *Data mining and knowledge discovery handbook* (pp. 53–82). Springer, Boston, MA.
5. Shlens, J. (2014). A tutorial on principal component analysis. *arXiv preprint arXiv: 1404.1100.*
6. Wall, M. E., Rechtsteiner, A., & Rocha, L. M. (2003). Singular value decomposition and principal component analysis. In *A practical approach to microarray data analysis* (pp. 91–109). Springer, Boston, MA.

3 Dual PCA

3.1 EXPLANATION AND WORKING

Dual Principal Component Analysis (PCA) is a variant of classical PCA. The goal is to reduce the dimensionality of the dataset but at the same time preserve maximum possible information from the original dataset.

- *Problem Statement:* Let the dataset contain a set of n data points denoted by $x_1, x_2, ..., x_n$ where each x_i is a d-dimensional vector. We try to find a p-dimensional linear subspace (where $p < d$, and often $p \ll d$) such that the data points lie mainly on this linear subspace. The linear subspace can be defined by p orthogonal vectors, say: $U_1, U_2, ..., U_p$. This linear subspace forms a new coordinate system.
- Let X be a $d \times n$ matrix which contains all the data points in the original space which has to be mapped to another $p \times n$ matrix Y (matrix) which retains maximum variability of the data points by reducing the number of features to represent the data point.

We saw that by using Singular Value Decomposition (SVD) on matrix X ($X = U D V^T$), the problem of eigendecomposition for PCA can be solved. This method works just perfectly when there are more data points than the dimensionality of the data (i.e., $d < n$). However, if $d > n$, the $d \times d$ matrix's eigendecomposition would be computationally expensive compared to the eigendecomposition of the $n \times n$ matrix. Figure 3.1 represents the case when $d > n$ [1, 2].

Hence, if we come up with a case where $d > n$, it would be beneficial for us to decompose $X^T X$ which is an $n \times n$ matrix, instead of decomposing XX^T which is a $d \times d$ matrix. This is where dual PCA comes into the picture.

Knowing PCA could be solved using SVD, we can compute U in terms of X, D, and V [3]. Since:

$$X = U D V^T \tag{3.1}$$

$$X V = U D V^T V \tag{3.2}$$

and V is an orthonormal matrix, $V^T V = 1$, and so:

$$X V = U D \tag{3.3}$$

$$U = X V D^{-1} \tag{3.4}$$

X U D V^T

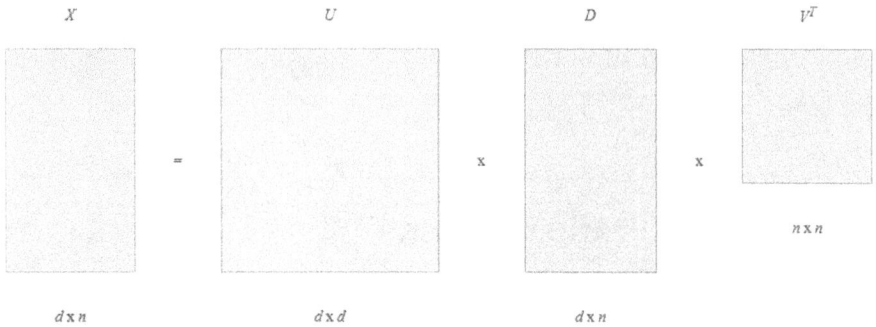

$d \times n$ $d \times d$ $d \times n$

FIGURE 3.1   Representation of singular value decomposition when $d > n$.

Replacing all use cases of U with $X V D^{-1}$ gives the dual form of PCA. Let us discuss the use cases that are most commonly used.

3.1.1   PROJECT DATA IN P-DIMENSIONAL SPACE

For any linear projection-based techniques, given X, we obtain a low dimensional representation Y such that

$$Y = U^T X \qquad (3.5)$$

We know that $X = U D V^T$. So, *multiplying* U^T on the left side we get

$$U^T X = U^T U D V^T \qquad (3.6)$$

Since U is an orthonormal matrix, $U^T U = 1$, so (3.6) becomes

$$U^T X = D V^T \qquad (3.7)$$

Since $Y = U^T X$, we get

$$Y = D V^T \qquad (3.8)$$

Hence, matrix X can be mapped to a lower dimensional subspace Y, just by using D and V.

3.1.2   RECONSTRUCT TRAINING DATA

We try to reconstruct the input matrix using the lower dimensional subspace Y such that

$$X' = U Y \qquad (3.9)$$

We know that, $U = X V D^{-1}$ and $Y = D V^T$. By substituting U with $X V D^{-1}$ and Y with $D V^T$ in the (3.9):

$$X' = X V D^{-1} D V^T \tag{3.10}$$

$$X' = X V V^T \tag{3.11}$$

Hence the lower dimensional data Y can be reconstructed back to the matrix X'.

3.1.3   PROJECT TEST DATA IN P-DIMENSIONAL SPACE

Let x be a d-dimensional test data point (out of the sample data) and we try to find its lower dimensional (p-dimensional) representation y:

$$y = U^T x \tag{3.12}$$

We know that, $U = X V D^{-1}$. By substituting U^T with $D^{-1} V^T X$:

$$y = D^{-1} V^T X x \tag{3.13}$$

It is possible to project a d-dimensional test data (out of sample) in a p-dimensional (lower dimensional) space.

3.1.4   RECONSTRUCT TEST DATA

We try to reconstruct the test data using the lower dimensional representation y (of the point x) such that

$$x' = U y \tag{3.14}$$

We know that $y = U^T x$. By substituting y with $U^T x$ in the (3.14), we get

$$x' = U U^T x \tag{3.15}$$

We know that $U = X V D^{-1}$. By substituting U with $X V D^{-1}$ and U^T with $D^{-1} V^T X^T$

$$x' = X V D^{-1} D^{-1} V^T X^T x \tag{3.16}$$

$$x' = X V D^{-2} V^T X^T x \tag{3.17}$$

Hence, test data x' can be reconstructed from the lower dimensional representation y.
    Dual PCA is a variant of PCA used when the number of features is greater than the number of data points. Since it is just a variant of PCA, it follows all the advantages and limitations of PCA.

REFERENCES

1. Ghodsi, A. (2006). Dimensionality reduction a short tutorial. *Department of Statistics and Actuarial Science, Univ. of Waterloo, Ontario, Canada*, *37*(38).
2. Ghojogh, B., & Crowley, M. (2019). Unsupervised and supervised principal component analysis: Tutorial. *arXiv preprint arXiv:1906.03148.*
3. Wall, M. E., Rechtsteiner, A., & Rocha, L. M. (2003). Singular value decomposition and principal component analysis. In *A practical approach to microarray data analysis*, 91–109. Springer, Boston, MA.

4 Kernel PCA

4.1 EXPLANATION AND WORKING

After understanding the concept of dimensionality reduction and a few algorithms for the same, let us now examine some plots given in Figure 4.1. What is the true dimensionality of these plots?

All dimensionality reduction techniques are based on the implicit assumption that the data lies along some low dimensional manifold. This is the case for the first three examples in Figure 4.1, which lie along a one-dimensional manifold even though it is plotted in a two-dimensional plane. In the fourth example in Figure 4.1, the data has been randomly plotted on a two-dimensional plane, so dimensionality reduction without losing information is not possible.

For the first two examples, we can use Principal Component Analysis (PCA) to find the approximate lower dimensional linear subspace. However, PCA will make no difference in the case of the third and fourth example because the structure is nonlinear and PCA only aims at finding the linear subspace. However, there are ways to find nonlinear lower dimensional manifolds.

Any form of linear projection to one dimension on this nonlinear data will result in linear principal components and we might lose information about the original dataset. This is because we need to consider nonlinear projection to one dimension to obtain the manifold on which the data points lie. So, how do we modify the PCA algorithm to solve for the nonlinear subspace in which the data points lie? In short, how do we make PCA nonlinear?

This is done using an idea similar to Support Vector Machines. Instead of using the original two-dimensional data points in one dimension using linear projections, we first write the data points as points in higher dimensional space. For example, say we write every two-dimensional point $x_t = (X_t, Y_t)$ into a 3-dimensional point given by mapping Φ as

$$\Phi(x_t) = (X_t, Y_t, X_t^2 + Y_t^2) \tag{4.1}$$

After this, instead of doing PCA on the original dataset, we perform PCA on $\Phi(x_1)$, $\Phi(x_2)$, ..., $\Phi(x_n)$. This process is known as Kernel PCA. So, the basic idea of Kernel PCA is to take the original data set and implicitly map it to a higher dimensional space using mapping Φ. Then we perform PCA on this space, which is linear projection in this higher dimensional space that already captures non-linearities in the original dataset [1, 2, 3].

4.1.1 KERNEL TRICK

Let us consider non-linear dimensionality reduction where the mapping Φ maps a vector x_t to a higher dimension that captures, say, all r^{th} order polynomials. For

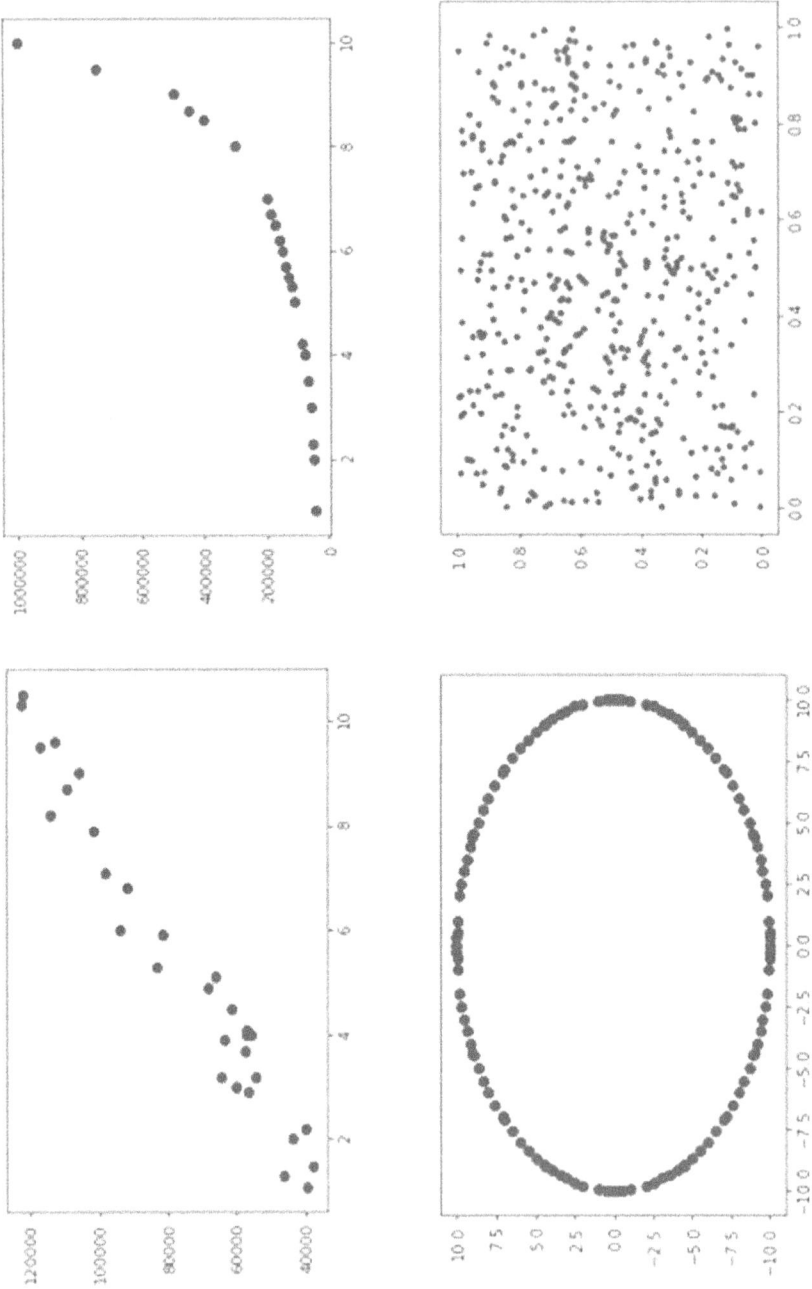

FIGURE 4.1 Plots of example data.

example, let us consider $x_t = (X_t, Y_t)$ is two-dimensional. And say we want to represent all third-degree polynomials, we need to consider:

$$\Phi(x_t) = \left(x_t^3, \; y_t^3, \; x_t^2 y_t, \; y_t^2 x_t, \; x_t^2, \; y_t^2, \; x_t y_t, \; x_t, \; y_t, 1 \right) \tag{4.2}$$

This mapped vector in the higher dimensional space is generally referred to as feature space.

Moreover, in many cases, we might have to map to infinite dimensions. In such a case, enumerating $\Phi(x_t)$'s is not computationally feasible. This is where the kernel trick comes handy. It works on a straightforward idea: The algorithm is modified so that it only needs to access the inner products between data points (because it is computationally feasible). And then, these inner products are replaced with the inner products in the feature space.

The kernel trick relies on the fact that while we might not be able to explicitly compute the mappings $\Phi(x_t)$, for any two x_t, x_s, we can compute the inner product using the kernel function as $k(x_t, x_s) = \Phi(x_t)^T \Phi(x_s)$ [4]. Let us understand this using an example.

Let's say $x_t = (X_t, Y_t)$ and p is some integer. Let's take the kernel function to be

$$k(x_t, x_s) = \left(x_t^T x_s \right)^p$$
$$= \left(X_t X_s + Y_t Y_s \right)^p \tag{4.3}$$

Using the binomial theorem on (4.3), we get:

$$k(x_t, x_s) = \sum_{i=1}^{p} (pC_i)(X_t X_s)^p (Y_t Y_s)^{p-i} \tag{4.4}$$

Now if we define $\Phi((X, Y)) = \left(X^p, \; pYX^{p-1}, \; \ldots, \; pC_i Y^i X^{p-i}, \; \ldots, \; Y^p \right)$, then

$$\Phi(x_t)^T \Phi(x_s) = \sum_{i=1}^{p} (pC_i)(X_t X_s)^p (Y_t Y_s)^{p-i} = k(x_t, x_s) \tag{4.5}$$

Now, to perform Kernel PCA, all we need to do is write PCA in such a way that it only depends on inner products.

Here, we rewrite PCA entirely in terms of inner products. We assume that the data in the feature space is centered. That is,

$$\frac{1}{n} \sum_{t=1}^{n} \Phi(x_t) = 0 \tag{4.6}$$

Let C be the covariance matrix:

$$C = \frac{1}{n} \sum_{t=1}^{n} \Phi(x_t) \Phi(x_t)^T \tag{4.7}$$

Let the eigenvectors be

$$C \, v_j = \lambda_j \, v_j \tag{4.8}$$

We want to avoid explicitly working with feature space. Instead, we want to work with kernels:

$$k(x_t, x_s) = \Phi(x_t)^T \Phi(x_s) \tag{4.9}$$

Rewriting the PCA equation:

$$\frac{1}{n}\sum_{t=1}^{n}\Phi(x_t)\Phi(x_t)^T v_j = \lambda_j v_j \tag{4.10}$$

Hence, the eigenvectors can be written as a linear combination of features:

$$v_j = \sum_{t=1}^{n}a_{jt}\Phi(x_t) \tag{4.11}$$

Finding the eigenvectors is equivalent to finding the coefficients a_{jt}.

Thus, now the equation becomes:

$$\frac{1}{n}\sum_{t=1}^{n}\Phi(x_t)\Phi(x_t)^T \sum_{t=1}^{n}a_{jt}\Phi(x_t) = \lambda_j \sum_{t=1}^{n}a_{jt}\Phi(x_t) \tag{4.12}$$

We can rewrite this as:

$$\frac{1}{n}\sum_{t=1}^{n}\Phi(x_t)\left(\sum_{i=1}^{n}a_{ji}\,k(x_t, x_i)\right) = \lambda_j \sum_{i=1}^{n}a_{ji}\Phi(x_i) \tag{4.13}$$

Multiplying by $\Phi(x_k)^T$ on the left side:

$$\frac{1}{n}\sum_{t=1}^{n}\Phi(x_k)^T \Phi(x_t)\left(\sum_{i=1}^{n}a_{ji}\,k(x_t, x_i)\right) = \lambda_j \sum_{i=1}^{n}a_{ji}\Phi(x_k)^T \Phi(x_i) \tag{4.14}$$

By plugging in the kernel again:

$$\frac{1}{n}\sum_{t=1}^{n}k(x_k, x_t)\left(\sum_{i=1}^{n}a_{ji}\,k(x_t, x_i)\right) = \lambda_j \sum_{i=1}^{n}a_{ji}k(x_k, x_i) \tag{4.15}$$

By rearranging the terms in (4.15), we get

$$k^2 a_j = n\,\lambda_j\,k\,a_j \tag{4.16}$$

We can remove a factor of k from both sides of the matrix (this will only affect eigenvectors with eigenvalues 0, which will not be principal components anyway):

$$k\,a_j = n\,\lambda_j\,a_j \tag{4.17}$$

We have a normalization condition for the a_j vectors:

$$v_j^T v_j = 1$$

$$\Rightarrow \sum_{k=1}^{n}\sum_{i=1}^{n}a_{ji}a_{jk}\Phi(x_i)^T \Phi(x_k) = 1 \tag{4.18}$$

$$\Rightarrow a_j^T\,k\,a_j = 1$$

Plugging this into $k\,a_j\,=\,n\,\lambda_j\,a_j$, we get

$$n\,\lambda_j\,a_j^T a_j\,=\,1,\,\forall j \tag{4.19}$$

For a new point x, its projection onto the principal components is

$$\Phi(x)^T v_j\,=\,\sum_{t=1}^{n} a_{jt}\Phi(x)^T\,\Phi(x_t)\,=\,\sum_{t=1}^{n} a_{jt}\,k\left(x,\,x_t\right) \tag{4.20}$$

Earlier in this section, we assumed that the feature space is centered. However, in general, the features $\Phi(x_t)$ may not have mean 0. We want to work with

$$\Phi(x_t)'\,=\,\Phi(x_t)\,-\,\frac{1}{n}\sum_{u=1}^{n}\Phi(x_u) \tag{4.21}$$

And the corresponding kernel matrix is given by

$$k\left(x_t,\,x_s\right)'\,=\,\left(\Phi(x_t)'\right)^T\,\Phi(x_s)' \tag{4.22}$$

$$k\left(x_t,\,x_s\right)'\,=\,\left(\Phi(x_t)\,-\,\frac{1}{n}\sum_{u=1}^{n}\Phi(x_u)\right)^T\left(\Phi(x_s)\,-\,\frac{1}{n}\sum_{u=1}^{n}\Phi(x_u)\right)$$

$$=\,\Phi(x_t)^T\Phi(x_s)\,-\,\left(\frac{1}{n}\sum_{u=1}^{n}\Phi(x_u)\right)^T\Phi(x_s)\,-\,\left(\frac{1}{n}\sum_{u=1}^{n}\Phi(x_u)\right)^T\Phi(x_t)$$

$$+\,\frac{1}{n^2}\left(\sum_{u=1}^{n}\Phi(x_u)\right)^T\left(\sum_{v=1}^{n}\Phi(x_v)\right)$$

$$=\,\Phi(x_t)^T\,\Phi(x_s)\,-\,\frac{1}{n}\sum_{u=1}^{n}(\Phi(x_u)\,)^T\,\Phi(x_s)\,-\,\frac{1}{n}\sum_{u=1}^{n}(\Phi(x_u)\,)^T\,\Phi(x_t)$$

$$+\,\frac{1}{n^2}\sum_{u=1}^{n}\sum_{v=1}^{n}(\Phi(x_u)\,)^T\,\Phi(x_v)$$

$$=\,k(t,\,s)\,-\,\frac{1}{n}\sum_{u=1}^{n}k(u,\,s)\,-\,\frac{1}{n}\sum_{u=1}^{n}k(u,\,t)\,+\,\frac{1}{n^2}\sum_{u=1}^{n}\sum_{v=1}^{n}k(u,\,v) \tag{4.23}$$

Here, $k(t,\,s)\,=\,\Phi(x_t)^T\,\Phi(x_s)$, where $\Phi(x_t)$ this time is not centered. Thus, it can be seen that the centered inner product matrix can be calculated based on the uncentered inner product matrix.

4.2 ADVANTAGES AND LIMITATIONS

4.2.1 KERNEL PCA VS. PCA

- Linear vs. nonlinear structure: kPCA can capture nonlinear structure in the data (if using a nonlinear kernel), whereas PCA cannot. However, if the data truly live on a linear manifold, kPCA cannot do better than

PCA. Using a nonlinear kernel may give worse performance due to overfitting.

- Inverse mapping: PCA allows inverse mapping from the lower dimensional space back to the original input space. So, the input points can be approximately reconstructed from their low dimensional representation. kPCA does not essentially provide an inverse mapping. However, it is still possible to estimate it using additional methods but at the cost of extra complexity and computational resources.

- Hyperparameters: We need to choose the number of dimensions for both PCA and kPCA initially. Apart from this, for kPCA, we also need to choose the kernel function and any associated parameters. This choice and the method employed to make this choice depend on the problem that we have on hand. Usually, we need to re-fit kPCA multiple times to compare different kernel and parameter choices.

- Computational cost: Usually, PCA has lower memory and runtime requirements than kPCA, and it could be scaled to massive datasets. There are many strategies for scaling up kPCA, but it requires making various approximations, the details of which are not in the scope of this chapter.

4.3   USE CASES

- Kernel PCA is one of the most commonly used manifold learning techniques. There is a range of practical applications of Kernel PCA. One of the simplest applications is the reconstruction of pre-images. Kernel PCA can be applied to improve traditional Active State Models (ASMs). Kernel PCA-based ASMs can be implemented and used to construct human face models [5].

- Another important application of Kernel PCA is using it as a nonlinear feature extractor as a preprocessing step for classification algorithms. Furthermore, it can also be considered as a generalization of linear PCA [4].

- Kernel PCA is one of the most popular techniques for feature extraction. It enables us to extract non-linear features and performs as a powerful preprocessing step for classification. However, there is one drawback. Extracted feature components are sensitive to outliers contained in data. This property is common to all PCA-based techniques. However, it is possible to remove outliers in data vectors and replace them with the estimated values via kernel PCA. By repeating this process several times, we can get the feature components less affected with outliers. This method can be applied for facial recognition tasks [6].

4.4   EXAMPLES AND TUTORIAL

To understand this method better, let us consider an example where we perform dimensionality reduction on high dimensional datasets using Kernel PCA and visualize the results.

Example 1

Let us see the results of dimensionality reduction on the 64-dimensional handwritten digits dataset (not to be confused with the MNIST handwritten digits database) consisting of 1083 images of six handwritten digits (0–5) each of size 8 × 8 pixels.
    Let us import the digits dataset from the sklearn library. Similar to the previous example, we will use sklearn.random_projection module from the Scikit-learn library for dimensionality reduction and matplotlib for plotting the data.
    First, import all the necessary libraries.

```
import sklearn
from sklearn import datasets
from sklearn.decomposition import KernelPCA
import matplotlib.pyplot as plt
```

Next, load the digits dataset using *load_digits*. This dataset originally contains 1797 data samples. That is, it contains 1797 images of digits. Each data point is an image of a digit of size 8 × 8. Hence, the dimensionality of the data is 64. The dataset is divided into 10 classes (digits from 0 to 9). Thus, the shape of the data is [1797, 64]. However, in this example we will consider only 6 classes (digits with label 0–5) that contain 1083 samples.

```
digits = datasets.load_digits(n_class=6)
X = digits.data
y = digits.target
n_samples, n_features = X.shape
print(n_features)
print(n_samples)
```

Output:

```
64
1083
```

Here, digits.data contains the images and digits.target contains the corresponding labels similar to mnist.train.images and mnist.train.labels respectively
    We use the Kernel PCA with a polynomial kernel. For this we use *sklearn. decomposition.KernelPCA*. The parameter *n_components* denotes the dimensionality of the target projection space and the parameter *kernel* denotes the kernel function that we will use. We then fit and transform the training data.

```
kpca =KernelPCA(n_components=2, kernel = 'poly')
X_kpca = kpca.fit_transform(X_train)
```

Now, plot the transformed data using matplotlib. Figure 4.2 visualizes the two-dimensional embedding obtained after performing dimensionality reduction on the 64-dimensional data consisting of 1083 data samples using Gaussian random projections.

```
plt.figure(figsize=(10,10))
plt.scatter(X_kpca[y==0, 0], X_ kpca[y==0, 1], color='blue',
alpha=0.5,label='0', s=9, lw=1)
```

```
plt.scatter(X_ kpca[y==1, 0], X_ kpca[y==1, 1], color='green',
alpha=0.5,label='1',s=9, lw=1)
plt.scatter(X_kpca[y==2, 0], X_kpca[y==2, 1], color='orange',
alpha=0.5,label='2', lw=1)
plt.scatter(X_kpca[y==3, 0], X_ kpca[y==3, 1], color='purple',
alpha=0.5,label='3',s=9, lw=1)
plt.scatter(X_ kpca[y==4, 0], X_ kpca[y==4, 1],
color='violet', alpha=0.5,label='4',s=9, lw=1)
plt.scatter(X_ kpca[y==5, 0], X_ kpca[y==5, 1], color='red',
alpha=0.5,label='5',s=9, lw=1)
plt.ylabel('Y coordinate')
plt.xlabel('X coordinate')
plt.legend()
plt.show()
```

Figure 4.2 shows the visualization of the digits data reduced to a two-dimensional feature space using Kernel PCA with polynomial kernel.

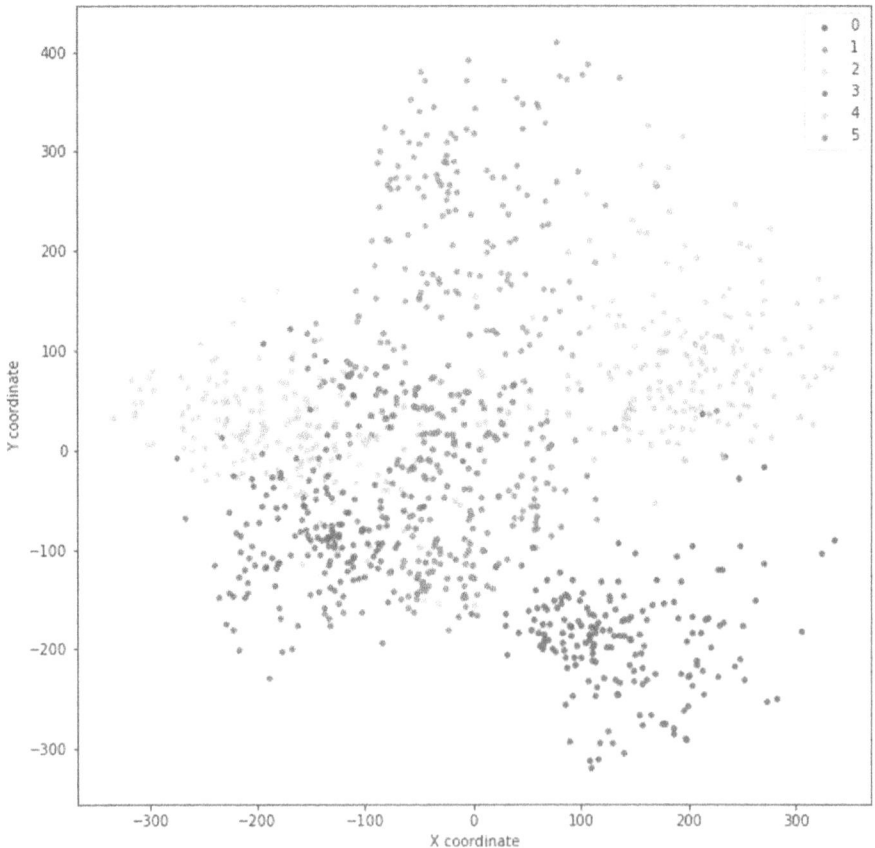

FIGURE 4.2 Kernel PCA on digits data.

Example 2

To demonstrate how Kernel PCA works as a manifold learning technique, let us consider an illustration of dimensionality reduction on an S-curve dataset. This toy dataset has 2000 data points lying on an S-shaped manifold. The objective of manifold learning is to learn the underlying intrinsic geometry of the manifold and unfold the manifold in low dimensional space by preserving the locality of the data points.

For this example, we use sklearn to create an S-curve dataset with 2000 data points. Likewise, we use *KernelPCA* algorithm implementation from the *sklearn.decomposition* module. Finally, to plot the data, we use matplotlib.

Firstly, import all the required libraries.

```
import sklearn
from sklearn.decomposition import KernelPCA
from sklearn import datasets
import matplotlib.pyplot as plt
```

Next, load the S-curve dataset with 2000 data points *(n_points)* using *datasets.make_s_curve*.

```
n_points = 2000
X, color = datasets.make_s_curve(n_points, random_state=0)
```

Now, create a 3D plot of the S-curve dataset.

```
fig = plt.figure(figsize=(45, 25))
ax = fig.add_subplot(251, projection='3d')
ax.scatter(X[:, 0], X[:, 1], X[:, 2], c=color, cmap=plt.
cm.jet, s=9, lw=1)
ax.view_init(10, -72)
```

Figure 4.3 is the visualization of the S-curve dataset on a three-dimensional scatter plot.

Next, we use Kernel PCA with rbf kernel and dimensionality of the target projection space *(n_components)* as 2.

```
kpca= KernelPCA(n_components=2, kernel = 'rbf')
X_kpca= kpca.fit_transform(X)
```

After fitting and transforming the data, plot the results.

```
fig = plt.figure(figsize=(5, 5))
ax = fig.add_subplot(1,1,1)
ax.scatter(X_pca[:, 0], X_pca[:, 1], c=color, cmap=plt.cm.jet,
s=9, lw=1)
ax.xaxis.set_major_formatter(NullFormatter())
ax.yaxis.set_major_formatter(NullFormatter())
ax.axis('tight')
plt.ylabel('Y coordinate')
plt.xlabel('X coordinate')
plt.show()
```

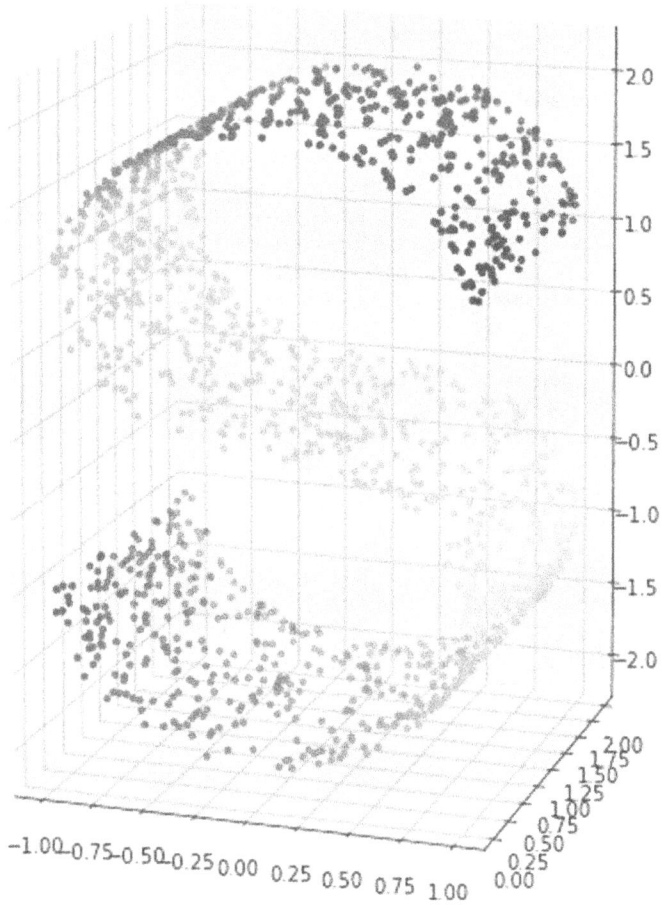

FIGURE 4.3   S-shaped curve, n=2000.

Figure 4.4 illustrates the result of manifold learning by Kernel PCA on the S-curve dataset with 2000 points.

Example 3

For this example, we use sklearn to create a Swiss roll dataset with 2000 data points. Likewise, we use *KernelPCA* algorithm implementation from the *sklearn,decomposition* module. Finally, to plot the data, we use matplotlib.
   Firstly, import all the required libraries.

```
from sklearn import datasets
from sklearn.decomposition import KernelPCA
import matplotlib.pyplot as plt
from matplotlib.ticker import NullFormatter
```

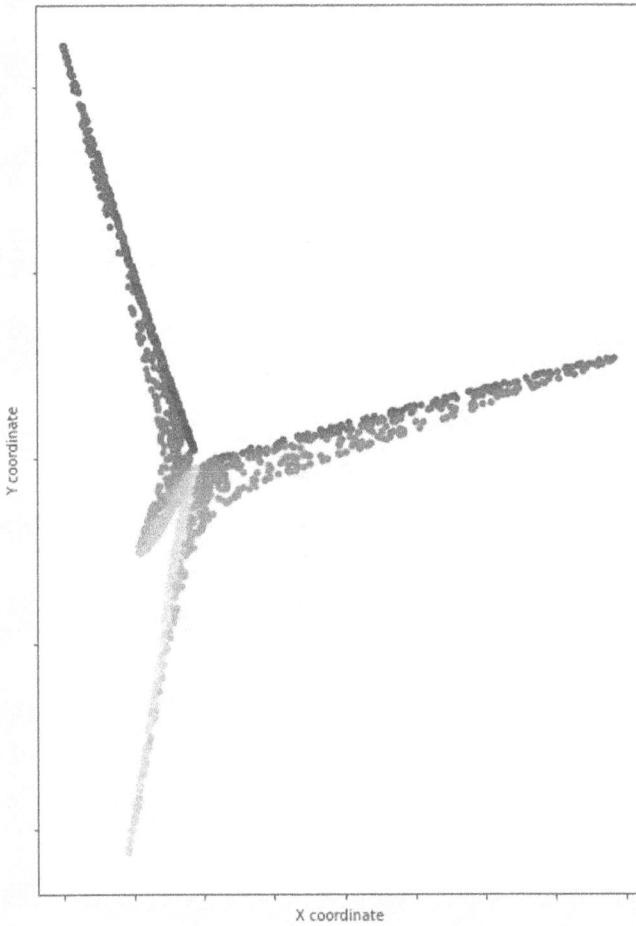

FIGURE 4.4 Kernel PCA.

Next, load the Swiss roll dataset with 2000 data points *(n_points)* using *datasets. make_swiss_roll.*

```
n_points = 2000
X, color = datasets.make_swiss_roll(n_points, random_state=0)
```

Now, create a 3D plot of the Swiss roll dataset.

```
fig = plt.figure(figsize=(45, 25))
ax = fig.add_subplot(251, projection='3d')
ax.scatter(X[:, 0], X[:, 1], X[:, 2], c=color, cmap=plt.
cm.jet, s=9, lw=1)
ax.view_init(10, -72)
```

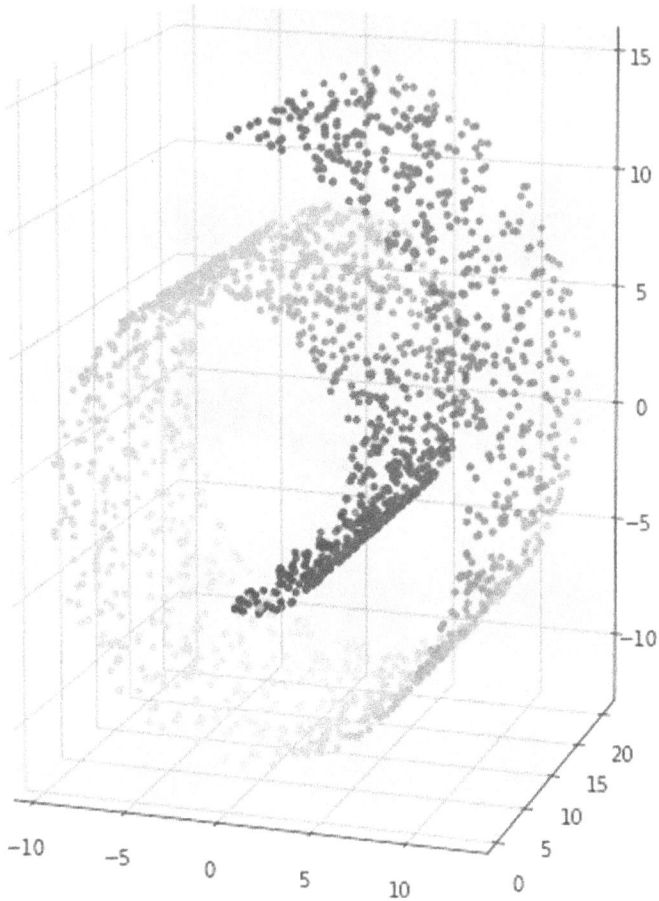

FIGURE 4.5 Swiss roll dataset, n=2000.

Figure 4.5 is the visualization of the Swiss roll dataset on a three-dimensional scatter plot.

Next, we use Kernel PCA with cosine kernel and dimensionality of the target projection space *(n_components)* as 2.

```
kpca = KernelPCA (n_components=2, kernel = 'cosine')
X_kpca = kpca.fit_transform(X)
```

After fitting and transforming the data, plot the results.

```
fig = plt.figure(figsize=(5, 5))
ax = fig.add_subplot(1,1,1)
ax.scatter(X_kpca[:, 0], X_kpca[:, 1], c=color, cmap=plt.
cm.jet, s=9, lw=1)
```

```
ax.xaxis.set_major_formatter(NullFormatter())
ax.yaxis.set_major_formatter(NullFormatter())
ax.axis('tight')
plt.ylabel('Y coordinate')
plt.xlabel('X coordinate')
plt.show()
```

Figure 4.6 illustrates the result of manifold learning by Kernel PCA on the Swiss roll dataset with 2000 points.

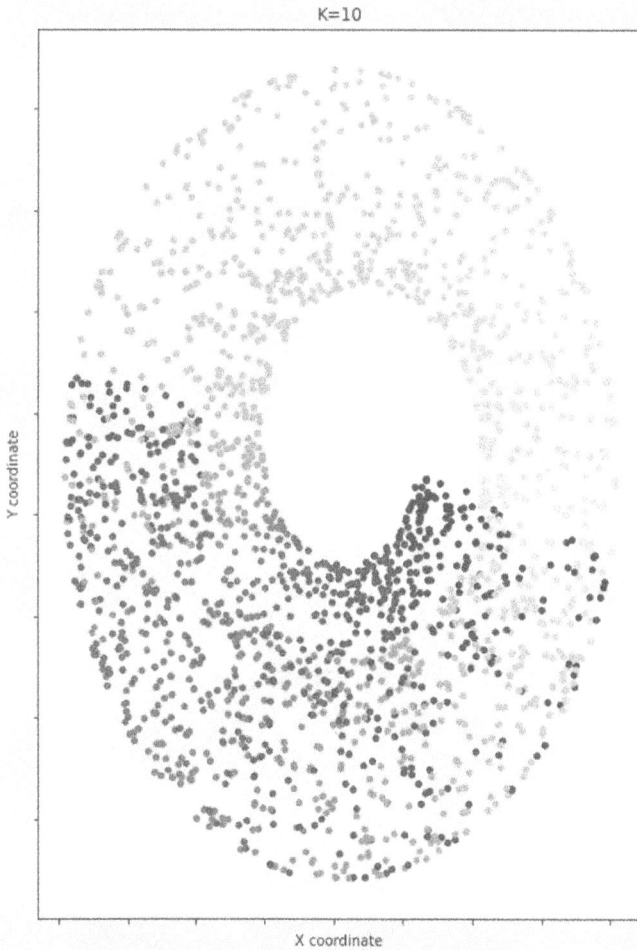

FIGURE 4.6 Kernel PCA on Swiss roll dataset.

REFERENCES

1. Ghojogh, B., & Crowley, M. (2019). Unsupervised and supervised principal component analysis: Tutorial. *arXiv preprint arXiv:1906.03148.*
2. Saul, L. K., Weinberger, K. Q., Sha, F., Ham, J., & Lee, D. D. (2006). Spectral methods for dimensionality reduction. *Semi-supervised learning*, *3*, pp. 293–308.
3. Burges, C. J. (2009). Geometric methods for feature extraction and dimensional reduction--a guided tour. In *Data mining and knowledge discovery handbook* (pp. 53–82). Springer, Boston, MA.
4. Mika, S., Schölkopf, B., Smola, A. J., Müller, K. R., Scholz, M., & Rätsch, G. (1999). Kernel PCA and de-noising in feature spaces. In *Advances in neural information processing systems* (pp. 536–542). MIT Press, Cambridge, MA.
5. Wang, Q. (2012). Kernel principal component analysis and its applications in face recognition and active shape models. *arXiv preprint arXiv:1207.3538.*
6. Takahashi, T., & Kurita, T. (2002). Robust de-noising by kernel PCA. In *International conference on artificial neural networks* (pp. 739–744). Springer, Berlin, Heidelberg.

5 Canonical Correlation Analysis (CCA)

5.1 EXPLANATION AND WORKING

Let us consider a scenario where we have to solve the task of speech recognition. That is, we have to convert the waveform received by the microphone to text. We are provided with a video of people speaking, so we have both visual and audio data. For speech recognition, we primarily rely on the audio data. However, apart from the voice of any person speaking, the audio data also consists of other unwanted noise, which hinders our task of speech recognition. Also, sometimes it might so happen that two or more people are speaking at one time, which leads to the superimposition of all the different waveforms generated by all the different people speaking at the same time. This also prevents our model from performing speech recognition efficiently. To improve the efficiency of the model in performing speech recognition in such cases, the visual data from the video that consists of relevant information for our task of speech recognition such as lip movements, facial expressions, and more can be used since the visual data might not contain information related to the background noise that is a part of the audio data.

In this scenario, we are interested in the information that is common to both the views. We use the visual data to filter out the noise and clean the audio data to get rid of the data specific only to the audio data. This type of scenario is precisely where techniques like Canonical Correlation Analysis (CCA) are useful [1].

To generalize, whenever we have two views of the same data points, and we are interested in the information in both the views, we are interested in the redundant information provided by both views. We can use this redundant information to reduce the noise in either one or both the views and obtain a lower dimensional representation of the data that consists of redundant information.

Now, let us move forward with the two-view compression problem. Let us say we are provided with two views of n data points in pairs: $\left(x_1, x_1'\right), \left(x_2, x_2'\right), ..., \left(x_n, x_n'\right)$ where each x_t is a d-dimensional vector and x_t' is a d'-dimensional vector. Here, x_n and x_t' are two views of the same data. Considering the scenario discussed above, say, if we have video data, x_t corresponds to visual data at some time t, and x_t' corresponds to audio data for the same point in time t. The goal is to compress these two views:

$$x_1, x_2, ..., x_n \; \forall x_i \in \mathbb{R}^d \rightarrow y_1, y_2, ..., y_n \; \forall y_i \in \mathbb{R}^p$$

and

$$x_1', x_2', ..., x_n' \; \forall x_i' \in \mathbb{R}^d \rightarrow y_1', y_2', ..., y_n' \; \forall y_i' \in \mathbb{R}^p$$

Moreover, we retain information that is common to both views. To do this, we use a linear transformation of data. Now, how do we find an appropriate linear transformation that retains as much redundant information as possible?

Let us begin by considering that we want to find a one-dimensional linear projection of both the views. That is we want to find $w_1 \in R^d$ and $v_1 \in R^{d'}$, where w_1 and v_1 are the first directions of CCA for two views of input data, so that $y_1, y_2, ..., y_n$ and $y_1', y_2', ..., y_n'$ retains maximum possible redundant information.

So, the first idea is to maximize the covariance between $y_1, y_2, ..., y_n$ and $y_1', y_2', ..., y_n'$. That is,

$$COV(Y,Y') = \frac{1}{n}\sum_{t=1}^{n}\left(y_t - \frac{1}{n}\sum_{s=1}^{n}y_s\right)\left(y_t' - \frac{1}{n}\sum_{s=1}^{n}y_s'\right)$$

(5.1)

However, there is an issue with this. The two views might be captured from entirely different sources. Say, for example, one source would be capturing distance and another source would be capturing amplitude. In such a case, the above idea would fail to solve our problem because maximizing covariance is scale-sensitive. So, if there are some coordinates that have high variance or scale, then this direction will dominate covariance.

So now, instead of maximizing covariance, we maximize covariance normalized by variance within each projection. This is commonly referred to as the correlation coefficient, and it is given by

$$Corr(Y,Y') = \frac{\frac{1}{n}\sum_{t=1}^{n}\left(y_t - \sum_{s=1}^{n}y_s\right)\left(y_t' - \frac{1}{n}\sum_{s=1}^{n}y_s'\right)}{\sqrt{\frac{1}{n}\sum_{t=1}^{n}\left(y_t - \frac{1}{n}\sum_{s=1}^{n}y_s\right)^2}\sqrt{\frac{1}{n}\sum_{t=1}^{n}\left(y_t' - \frac{1}{n}\sum_{s=1}^{n}y_s'\right)^2}}$$

(5.2)

We know that y_t and y_t' scale linearly with w_1 and v_1. Hence, if we scale w_1 to aw_1 then y_t also scales by a. Furthermore, the correlation coefficient is not affected by scaling y_t and y_t' as the a-scaling in the numerator is cancelled out by the denominator. So, we can appropriately scale w_1 and v_1 in such a way that

$$\frac{1}{n}\sum_{t=1}^{n}\left(y_t - \frac{1}{n}\sum_{s=1}^{n}y_s\right)^2 = \frac{1}{n}\sum_{t=1}^{n}\left(y_t' - \frac{1}{n}\sum_{s=1}^{n}y_s'\right)^2 = 1$$ and hence we arrive at a well-

defined optimization problem for finding the first directions of CCA (w_1 and v_1):

$$\max \frac{1}{n}\sum_{t=1}^{n}\left(y_t - \frac{1}{n}\sum_{s=1}^{n}y_s\right)\left(y_t' - \frac{1}{n}\sum_{s=1}^{n}y_s'\right)$$

$$\text{s.t. } \frac{1}{n}\sum_{t=1}^{n}\left(y_t - \frac{1}{n}\sum_{s=1}^{n}y_s\right)^2 = \frac{1}{n}\sum_{t=1}^{n}\left(y_t' - \frac{1}{n}\sum_{s=1}^{n}y_s'\right)^2 = 1$$

We can write this explicitly as:

$$w_1, v_1 = \arg\max \frac{1}{n}\sum_{t=1}^{n}\left(y_t - \frac{1}{n}\sum_{s=1}^{n}y_s\right)\left(y_t' - \frac{1}{n}\sum_{s=1}^{n}y_s'\right)$$

$$\text{s.t } \frac{1}{n}\sum_{t=1}^{n}\left(y_t - \frac{1}{n}\sum_{s=1}^{n}y_s\right)^2 = \frac{1}{n}\sum_{t=1}^{n}\left(y_t' - \frac{1}{n}\sum_{s=1}^{n}y_s'\right)^2 = 1 \qquad (5.3)$$

Equation (5.3) can be re-written as:

$$w_1, v_1 = \arg\max \frac{1}{n}\sum_{t=1}^{n}\left(w_1^T x_t - \frac{1}{n}w_1^T\left(\sum_{s=1}^{n}x_t\right)\right)\left(v_1^T x_t' - \frac{1}{n}v_1^T\left(\sum_{s=1}^{n}x_t'\right)\right)$$

$$\text{s.t } \frac{1}{n}\sum_{t=1}^{n}\left(w_1^T x_t - \frac{1}{n}w_1^T\left(\sum_{s=1}^{n}x_t\right)\right)^2 = \frac{1}{n}\sum_{t=1}^{n}\left(v_1^T x_t' - \frac{1}{n}v_1^T\left(\sum_{s=1}^{n}x_t'\right)\right)^2 = 1 \quad (5.4)$$

Now, let us consider $\mu = \frac{1}{n}\sum_{s=1}^{n}x_t$ and $\mu' = \frac{1}{n}\sum_{s=1}^{n}x_t'$. Rewriting (5.4):

$$w_1, v_1 = \arg\max \frac{1}{n}\sum_{t=1}^{n}\left(w_1^T(x_t - \mu)\right)\left(v_1^T(x_t' - \mu')\right)$$

$$\text{s.t } \frac{1}{n}\sum_{t=1}^{n}\left(w_1^T(x_t - \mu)\right)^2 = \frac{1}{n}\sum_{t=1}^{n}\left(v_1^T(x_t' - \mu')\right)^2 = 1 \qquad (5.5)$$

We know that $(a^T b)(c^T d) = a^T(bc^T)d$ and $c^T d = d^T c$. Hence, we can write

$$(a^T b)(c^T d) = a^T(bd^T)c$$

This can be written as:

$$w_1, v_1 = \arg\max \frac{1}{n}\sum_{t=1}^{n}w_1^T\left((x_t - \mu)(x_t' - \mu')^T\right)v_1$$

$$\text{s.t. } \frac{1}{n}\sum_{t=1}^{n}w_1^T\left((x_t - \mu)(x_t' - \mu')^T\right)w_1 = \frac{1}{n}\sum_{t=1}^{n}v_1^T\left((x_t - \mu)(x_t' - \mu')^T\right)v_1 = 1 \quad (5.6)$$

Thus,

$$w_1, v_1 = \arg\max \ w_1^T\left(\frac{1}{n}\sum_{t=1}^{n}\left((x_t - \mu)(x_t' - \mu')^T\right)\right)v_1$$

$$\text{s.t. } w_1^T\left(\frac{1}{n}\sum_{t=1}^{n}\left((x_t - \mu)(x_t' - \mu')^T\right)\right)w_1 = v_1^T\left(\frac{1}{n}\sum_{t=1}^{n}\left((x_t - \mu)(x_t' - \mu')^T\right)\right)v_1 = 1 \quad (5.7)$$

Now, $\Sigma_{1,1} = \dfrac{1}{n}\sum\limits_{t=1}^{n}\left((x_t - \mu)(x_t' - \mu')^T\right)$ is the covariance matrix for view 1;

$\Sigma_{2,2} = \dfrac{1}{n}\sum\limits_{t=1}^{n}\left((x_t - \mu)(x_t' - \mu')^T\right)$ is the covariance matrix for view 2; and

and $\Sigma_{1,2} = \dfrac{1}{n}\sum\limits_{t=1}^{n}\left((x_t - \mu)(x_t' - \mu')^T\right)$ is the covariance matrix between views

1 and 2.

Hence, the optimization problem can be written as:

$$w_1, v_1 = \arg\max\ w_1^T \Sigma_{1,2} v_1$$

$$\text{s.t. } w_1^T \Sigma_{1,1} w_1 = v_1^T \Sigma_{2,2} v_1 = 1 \tag{5.8}$$

This is the simplified optimization problem. To solve this optimization problem, we use the concept of Lagrange's multipliers:

$$w_1, v_1 = \arg\max\ w_1^T \Sigma_{1,2} v_1 + \lambda_1\left(1 - w_1^T \Sigma_{1,1} w_1\right) + \lambda_2\left(1 - v_1^T \Sigma_{2,2} v_1\right) \tag{5.9}$$

To solve (5.9), we take derivatives on both sides of the equation and equate it to 0:

$$\Sigma_{1,2} v_1 = \lambda_1 \Sigma_{1,1} w_1 \text{ and } \Sigma_{2,1} w_1 = \lambda_2 \Sigma_{2,2} v_1 \tag{5.10}$$

However, we know that $w_1^T \Sigma_{1,1} w_1 = 1$. Multiplying (5.10) by w_1^T on the left side:

$$w_1^T \Sigma_{1,2}\ v_1 = \lambda_1 \tag{5.11}$$

Similarly, for the other equation:

$$v_1^T \Sigma_{2,1}\ w_1 = \lambda_2 \tag{5.12}$$

So, $\lambda_1 = \lambda_2 = w_1^T \Sigma_{1,2}\ v_1 = v_1^T \Sigma_{2,1}\ w_1 = \lambda$

Now, rewriting $\Sigma_{2,1} w_1 = \lambda\ \Sigma_{2,2} w_1$ as $\Sigma_{2,2}^{-1}\Sigma_{2,1}\ w_1 = \lambda\ v_1$ and substituting it in $\Sigma_{1,2} v_1 = \lambda\ \Sigma_{1,1} w_1$:

$$\Sigma_{1,2}\ \Sigma_{2,2}^{-1}\ \Sigma_{2,1}\ w_1 = \lambda^2\ \Sigma_{1,1} w_1 \tag{5.13}$$

Multiplying both sides by $\Sigma_{1,1}^{-1}$:

$$\Sigma_{1,1}^{-1}\Sigma_{1,2}\ \Sigma_{2,2}^{-1}\ \Sigma_{2,1}\ w_1 = \lambda^2\ w_1 \tag{5.14}$$

Thus, w_1 is the eigenvector of $\Sigma_{1,1}^{-1}\Sigma_{1,2}\ \Sigma_{2,2}^{-1}\ \Sigma_{2,1}$. Furthermore, the objective is to maximize $w_1^T \Sigma_{1,2}\ v_1 = \lambda$. To maximize the objective, we need to maximize the eigenvalue of w_1. So, w_1 is the top eigenvector of $\Sigma_{1,1}^{-1}\Sigma_{1,2}\,\Sigma_{2,2}^{-1}\,\Sigma_{2,1}$. Similarly, v_1 is the top eigenvector of $\Sigma_{2,2}^{-1}\Sigma_{2,1}\,\Sigma_{1,1}^{-1}\,\Sigma_{1,2}$.

To find the remaining directions, we look for the next w_i and v_i that maximizes the objective function such that v_i is orthogonal to $v_1, v_2, ...,v_{i-1}$ and w_i is orthogonal to $w_1, w_2, ..., w_{i-1}$. Thus, W and V are the top p eigenvectors of $\Sigma_{1,1}^{-1}\Sigma_{1,2}\,\Sigma_{2,2}^{-1}\,\Sigma_{2,1}$ and $\Sigma_{2,2}^{-1}\Sigma_{2,1}\,\Sigma_{1,1}^{-1}\,\Sigma_{1,2}$ respectively.

So, the final solution to CCA is

$$W = eigs\left(\Sigma_{1,1}^{-1}\Sigma_{1,2}\,\Sigma_{2,2}^{-1}\,\Sigma_{2,1},\, p\right) \tag{5.15}$$

$$V = eigs\left(\Sigma_{2,2}^{-1}\Sigma_{2,1}\,\Sigma_{1,1}^{-1}\,\Sigma_{1,2},\, p\right) \tag{5.16}$$

5.2 ADVANTAGES AND LIMITATIONS OF CCA

- The solution of CCA reflects the variance shared by the linear composites of the sets of variables and not the variance extracted from these variables.
- CCA tries to maximize the correlation between linear composites, not to maximize the variance extracted.
- CCA is a useful technique for gaining insight into what otherwise may be an unmanageable number of bivariate correlations between sets of variables.
- CCA is a technique that can define structure in both the dependent and independent variates simultaneously. So, in situations where a series of measures are used for both dependent and independent variates, applying CCA is a logical choice.
- However, there are some drawbacks of CCA. Procedures that maximize the correlation do not necessarily maximize the interpretation of the pairs of canonical variates; therefore, canonical solutions are not easily interpretable.
- CCA is a multivariate technique, and so it limits the probability of committing a Type I error.

5.3 USE CASES AND EXAMPLES

- The speech recognition example is one of the most common use cases of CCA.
- Another case where CCA is of practical use is when we have multiple choices for feature extraction techniques, and all of these techniques are believed to work well. We could just concatenate all the extracted features from different feature extraction techniques to form a single feature in this situation. However, if these individual features are good enough for the task, then in that case, we can apply CCA to extract the redundant data among these features and, in the process, reduce noise that is individual to each of the feature extraction processes.
- Another typical use of CCA is in psychological testing, where we could take two well-established multidimensional personality tests, like the Minnesota Multiphasic Personality Inventory (MMPI-2) and the NEO. By

seeing how the MMPI-2 factors relate to the NEO factors, one could gain insight into what dimensions were shared between the tests and how much variance was shared. For example, we might find an extraversion or neuroticism dimension accounted for a substantial amount of shared variance between the two tests [2].

- We can also use CCA to produce a model equation which relates two sets of variables. For instance, let us take a set of performance measures and a set of explanatory variables, or say a set of outputs and inputs. Constraint restrictions can be imposed on such a model to ensure it reflects theoretical requirements or intuitively obvious conditions. This type of model is known as the maximum correlation model [2].

REFERENCES

1. Andrew, G., Arora, R., Bilmes, J., & Livescu, K. (2013, May). Deep canonical correlation analysis. In *International conference on machine learning* (pp. 1247–1255). PMLR, Atlanta, GA.
2. Hardoon, D. R., Szedmak, S., & Shawe-Taylor, J. (2004). Canonical correlation analysis: An overview with application to learning methods. *Neural computation*, *16*(12), 2639–2664.

6 Multidimensional Scaling (MDS)

6.1 EXPLANATION AND WORKING

Manifold learning is an approach for non-linear dimensionality reduction. That is, all manifold learning techniques assume that the data lies on a smooth, non-linear manifold of lower dimension and that a mapping $f : R^d \rightarrow R^p \ (d \gg p)$ can be found by preserving one or more properties of the higher dimensional space. Multidimensional scaling is a manifold learning technique [1]. More particularly, it is a distance-preserving manifold learning technique. Distance-preserving techniques assume that the manifold can be defined by the pairwise distance between the points on the manifold. Any distance-preserving manifold learning technique finds an underlying lower dimensional non-linear manifold by finding an appropriate mapping $f : R^d \rightarrow R^p \ (d \gg p)$ that preserves the pairwise distance between the points of the higher dimensional space [2, 3].

Distance-preserving techniques can be broadly classified into two classes:

1. Techniques that preserve spatial distance.
2. Techniques that preserve graphical distance.

Multidimensional Scaling (MDS) is a spatial distance-preserving manifold-learning technique. That is, it attempts to find a lower dimensional non-linear manifold that represents the data points such that it preserves the spatial distance between the points in the higher dimensional space.

As discussed above, it is important to maintain the pairwise distance between the points while reducing the dimensionality. Let a dataset consist of **n** data points denoted by $x_1, x_2, ..., x_n$ where each x_i is a **d**-dimensional vector. Let M be an n × n matrix that represents the pairwise distance between all n points, where $M_{i,j}$ is the distance between the ith and j^{th} point. We try to find $y_1, y_2, ..., y_n$ where each y_i is a **p**-dimensional vector (where **p** < **d**, and often **p** ≪ **d**) such that:

$$\left\| \, y_i - y_j \, \right\|_2 \approx M_{i, j}$$

First, let us define a matrix D (n × n matrix) such that for any $i, \ j$:

$$D_{i, j} = M_{i, j}^2 \tag{6.1}$$

That is, D is the entry-wise square of matrix M.

Now, let us consider that there exist points y_1', y_2', ..., y_n' such that their pairwise distances are equal to the distances represented in the matrix M. So, for any i, j:

$$D_{i,j} = \left\| y_i' - y_j' \right\|_2^2 = \left\| y_i' \right\|_2^2 + \left\| y_j' \right\|_2^2 - 2\left(y_i'\right)^T y_j' \tag{6.2}$$

This can be rewritten as:

$$\left(y_i'\right)^T y_j' = \frac{1}{2}\left(\left\| y_i' \right\|_2^2 + \left\| y_j' \right\|_2^2 - D_{i,j} \right) \tag{6.3}$$

Now, if we center these n points, their pairwise distance still remains the same. Hence, let us assume that $\sum_{i=1}^{n} y_i' = 0$. Using this:

$$\frac{1}{n}\sum_{i=1}^{n} D_{i,j} = \frac{1}{n}\sum_{i=1}^{n}\left(\left\| y_i' \right\|_2^2 \right) + \left\| y_j' \right\|_2^2 - \frac{2}{n}\sum_{i=1}^{n}\left(y_i'\right)^T y_j'$$

$$= \frac{1}{n}\Sigma_{i=1}^{n}\left(\left\| y_i' \right\|_2^2 \right) + \left\| y_j' \right\|_2^2 - \frac{2}{n}\left(\Sigma_{i=1}^{n} y_i'\right)^T y_j' \tag{6.4}$$

$$= \frac{1}{n}\Sigma_{i=1}^{n}\left(\left\| y_i' \right\|_2^2 \right) + \left\| y_j' \right\|_2^2$$

Similarly:

$$\frac{1}{n}\sum_{j=1}^{n} D_{i,j} = \frac{1}{n}\sum_{j=1}^{n}\left(\left\| y_j' \right\|_2^2 \right) + \left\| y_i' \right\|_2^2 \tag{6.5}$$

Furthermore:

$$\frac{1}{n^2}\sum_{i=1}^{n}\sum_{j=1}^{n} D_{i,j} = \frac{2}{n}\sum_{i=1}^{n}\left\| y_i' \right\|_2^2 \tag{6.6}$$

So, using (6.6) in

$$\left(y_i'\right)^T y_j' = \frac{1}{2}\left(\left\| y_i' \right\|_2^2 + \left\| y_j' \right\|_2^2 - D_{i,j} \right) \tag{6.7}$$

We have:

$$\left(y_i'\right)^T y_j' = \frac{1}{2}\left(\frac{1}{n}\sum_{i=1}^{n} D_{i,j} + \frac{1}{n}\sum_{j=1}^{n} D_{i,j} - \frac{1}{n^2}\sum_{i=1}^{n}\sum_{j=1}^{n} D_{i,j} - D_{i,j} \right) \tag{6.8}$$

The equation (6.8) can be succinctly written as:

$$Y'(Y')^T = \frac{1}{2}\left(\frac{1}{n}11^T D + \frac{1}{n}D11^T - \frac{1}{n^2}11^T D11^T - D \right)$$

$$= -\frac{1}{2}\left(I - \frac{1}{n}11^T \right)D\left(I - \frac{1}{n}11^T \right) \tag{6.9}$$

Here, the matrix Y' is a matrix with n rows, where $\left(y_t'\right)^T$ are the rows. Hence, it can be understood that $\left(I - \frac{1}{n} 1\,1^T\right) D \left(I - \frac{1}{n} 1\,1^T\right)$ is a matrix of inner products of centered data points (similar to what we discussed in kernel PCA). So, we exactly know how to obtain the low dimensional representation of these n points. The top p eigenvectors of $\left(I - \frac{1}{n} 1\,1^T\right) D \left(I - \frac{1}{n} 1\,1^T\right)$ matrix have to be found and these eigenvectors have to be scaled with the square root of the corresponding eigenvalues. This gives the final $n \times p$ matrix Y.

So, the final algorithm is:

1. Compute the matrix D that is the entry-wise square matrix of the given matrix M.
2. Compute $-\frac{1}{2} \left(I - \frac{1}{n} 1\,1^T\right) D \left(I - \frac{1}{n} 1\,1^T\right)$.
3. Compute $U_1, U_2, ..., U_p$, the top p eigenvectors with the corresponding eigenvalues $\lambda_1, \lambda_2, ..., \lambda_p$.
4. Compute the $n \times p$ matrix Y by setting the ith column to $\sqrt{\lambda_i}\, U_i$.

6.2   ADVANTAGES AND LIMITATIONS

- MDS is a distance-preserving manifold learning technique, and it tries to find an underlying non-linear lower dimensional subspace.
- MDS tries to maintain Euclidean distance in the lower dimension. However, the Euclidean metric for distance works only if the neighborhood structure can be approximated as a linear structure in non-linear manifolds.
- MDS requires large computing power for calculating the dissimilarity matrix at every iteration. It is hard to embed the new data in MDS.
- The outcome of an MDS is more dependent on the decisions that are taken beforehand. At the data collection stage, asking for similarity rather than for dissimilarity ratings might affect the results. For example, a similarity judgment cannot simply be regarded as the "inverse" of a dissimilarity judgment.

6.3   USE CASES

- MDS can be applied to an extensive range of datasets. Technically, some type of multidimensional scaling method can be used to analyze any matrix of transformed or raw data if the elements of the data matrix indicate the strength of relation between the objects or events represented by the rows and columns of the data matrix. Such data is called relational data with examples such as correlations, distances, proximities, similarities, multiple rating scales, preference matrices, etc.
- MDS methods are useful for all relational data matrices, including asymmetric and symmetric matrices, rectangular and square matrices, matrices

with or without missing elements, equally and unequally replicated data matrices, two-way and multi-way matrices, and other types of matrices [4].

- MDS is very popular and developed in the field of human sciences like sociology, anthropology, and especially in psychometrics.

To understand this method better, let us consider an example where we perform dimensionality reduction on high dimensional datasets using MDS and visualize the results.

6.4   EXAMPLES AND TUTORIAL

Example 1

To demonstrate how MDS works as a manifold learning technique, let us consider an illustration of dimensionality reduction on an S-curve dataset. This toy dataset has 2000 data points lying on an S-shaped manifold. The objective of manifold learning is to learn the underlying intrinsic geometry of the manifold and unfold the manifold in low dimensional space by preserving the locality of the data points.

For this example, we use sklearn to create an S-curve dataset with 2000 data points. Likewise, we use MDS algorithm implementation from the *sklearn, manifold* module. Finally, to plot the data, we use matplotlib.

Firstly, import all the required libraries.

```
from sklearn.manifold import MDS
from sklearn import datasets
import matplotlib.pyplot as plt
from matplotlib.ticker import NullFormatter
```

Next, load the S-curve dataset with 2000 data points *(n_points)* using *datasets. make_s_curve.*

```
n_points = 2000
X, color = datasets.make_s_curve(n_points, random_state=0)
```

Now, create a 3D plot of the S-curve dataset.

```
fig = plt.figure(figsize=(45, 25))
ax = fig.add_subplot(251, projection='3d')
ax.scatter(X[:, 0], X[:, 1], X[:, 2], c=color, cmap=plt.
cm.jet, s=9, lw=1)
ax.view_init(10, -72)
```

Figure 6.1 is the visualization of the S-curve dataset on a three-dimensional scatter plot.

Next, we use MDS with dimensionality of the target projection space *(n_components)* as 2.

```
mds= MDS (n_components=2)
X_mds= mds.fit_transform(X)
```

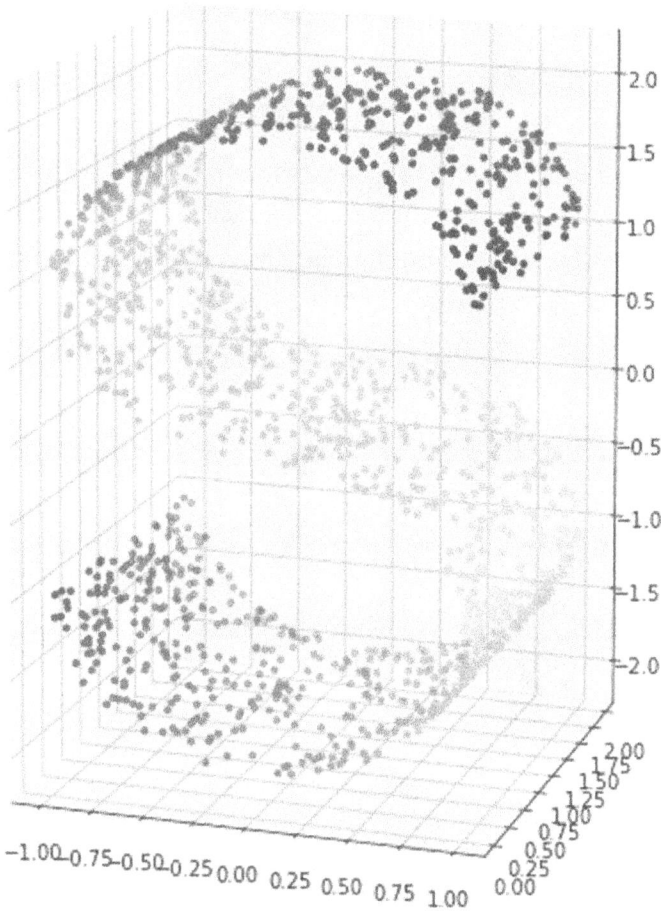

FIGURE 6.1 S-shaped curve, n=2000.

After fitting and transforming the data, plot the results.

```
fig = plt.figure(figsize=(5, 5))
ax = fig.add_subplot(1,1,1)
ax.scatter(X_mds[:, 0], X_mds[:, 1], c=color, cmap=plt.cm.jet,
s=9, lw=1)
ax.xaxis.set_major_formatter(NullFormatter())
ax.yaxis.set_major_formatter(NullFormatter())
ax.axis('tight')
plt.ylabel('Y coordinate')
plt.xlabel('X coordinate')
plt.show()
```

Figure 6.2 illustrates the result of manifold learning by MDS on the S-curve dataset with 2000 points.

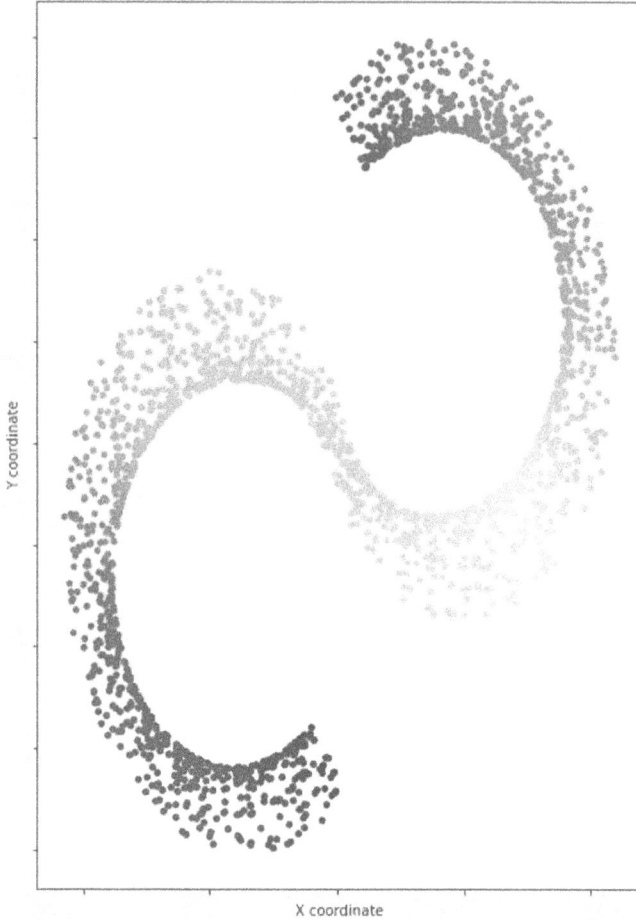

FIGURE 6.2 Multidimensional scaling.

Example 2

For this example, we use sklearn to create a Swiss roll dataset with 2000 data points. Likewise, we use MDS algorithm implementation from the *sklearn, manifold* module. Finally, to plot the data, we use matplotlib.

Firstly, import all the required libraries.

```
from sklearn.manifold import MDS
from sklearn import datasets
import matplotlib.pyplot as plt
from matplotlib.ticker import NullFormatter
```

Next, load the Swiss roll dataset with 2000 data points *(n_points)* using *datasets. make_swiss_roll.*

```
n_points = 2000
X, color = datasets.make_swiss_roll(n_points, random_state=0)
```

Now, create a 3D plot of the Swiss roll dataset.

```
fig = plt.figure(figsize=(45, 25))
ax = fig.add_subplot(251, projection='3d')
ax.scatter(X[:, 0], X[:, 1], X[:, 2], c=color, cmap=plt.
cm.jet, s=9, lw=1)
ax.view_init(10, -72)
```

Figure 6.3 is the visualization of the Swiss roll dataset on a three-dimensional scatter plot.

Next, we use MDS with dimensionality of the target projection space *(n_components)* as 2.

```
mds= MDS (n_components=2)
X_mds= mds.fit_transform(X)
```

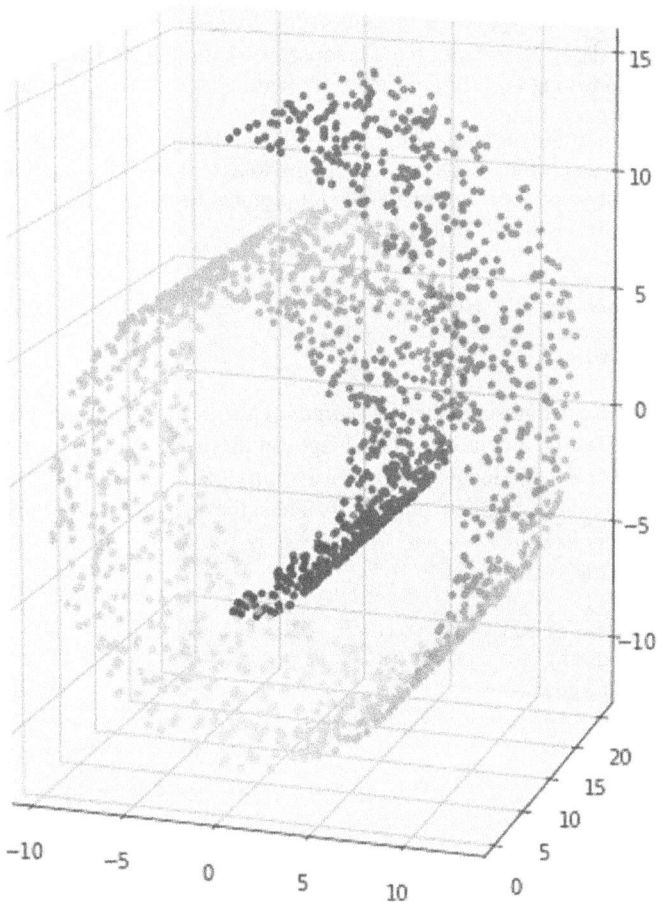

FIGURE 6.3   Swiss roll dataset, n=2000.

After fitting and transforming the data, plot the results.

```
fig = plt.figure(figsize=(5, 5))
ax = fig.add_subplot(1,1,1)
ax.scatter(X_mds[:, 0], X_mds[:, 1], c=color, cmap=plt.cm.jet,
s=9, lw=1)
ax.xaxis.set_major_formatter(NullFormatter())
ax.yaxis.set_major_formatter(NullFormatter())
ax.axis('tight')
plt.ylabel('Y coordinate')
plt.xlabel('X coordinate')
plt.show()
```

Figure 6.4 illustrates the result of manifold learning by MDS on the Swiss roll data-set with 2000 points.

Example 3

Next, let us see the results of dimensionality reduction on the 64-dimensional handwritten digits dataset (not to be confused with the MNIST handwritten digits database) consisting of 1083 images of six handwritten digits (0–5) each of size 8 × 8 pixels.

Let us import the digits dataset from the sklearn library. Similar to the previous example, we will use the sklearn.manifold module from the Scikit-learn library for dimensionality reduction and matplotlib for plotting the data.

First, import all the necessary libraries.

```
import sklearn
from sklearn import datasets
from sklearn.manifold import MDS
import matplotlib.pyplot as plt
```

Next, load the digits dataset using *load_digits*. This dataset originally contains 1797 data samples. That is, it contains 1797 images of digits. Each data point is an image of a digit of size 8 × 8. Hence, the dimensionality of the data is 64. The dataset is divided into 10 classes (digits from 0 to 9). Thus, the shape of the data is [1797, 64]. However, in this example we will consider only 6 classes (digits with label 0–5) which contain 1083 samples.

```
digits = datasets.load_digits(n_class=6)
X = digits.data
y = digits.target
n_samples, n_features = X.shape
print(n_features)
print(n_samples)
```

Output:

```
64
1083
```

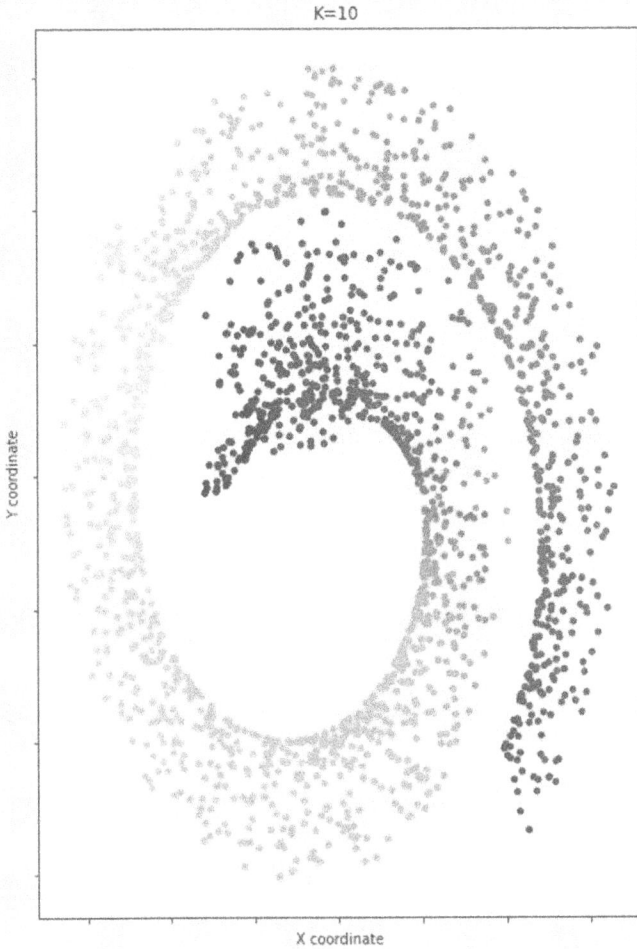

FIGURE 6.4   MDS on Swiss roll dataset.

Here, digits.data contains the images and digits.target contains the corresponding labels similar to mnist.train.images and mnist.train.labels, respectively.

We use *sklearn.manifold.MDS* for MDS. The parameter *n_components* denotes the dimensionality of the target projection space. We then fit and transform the training data.

```
mds = MDS(n_components=2)
X_mds = mds.fit_transform(X)
```

Now, plot the transformed data using matplotlib..

```
plt.figure(figsize=(10,10))
plt.scatter(X_mds[y==0, 0], X_mds[y==0, 1], color='blue',
alpha=0.5,label='0', s=9, lw=2)
```

```
plt.scatter(X_mds[y==1, 0], X_mds[y==1, 1], color='green',
alpha=0.5,label='1',s=9, lw=2)
plt.scatter(X_mds[y==2, 0], X_mds[y==2, 1], color='orange',
alpha=0.5,label='2',s=9, lw=2)
plt.scatter(X_mds[y==3, 0], X_mds[y==3, 1], color='purple',
alpha=0.5,label='3',s=9, lw=2)
plt.scatter(X_mds[y==4, 0], X_mds[y==4, 1], color='violet',
alpha=0.5,label='4',s=9, lw=2)
plt.scatter(X_mds[y==5, 0], X_mds[y==5, 1], color='red',
alpha=0.5,label='5',s=9, lw=2)
plt.ylabel('Y coordinate')
plt.xlabel('X coordinate')
plt.legend()
plt.show()
```

Figure 6.5 visualizes the two-dimensional embedding obtained after performing dimensionality reduction on the 64-dimensional data consisting of 1083 data samples using MDS.

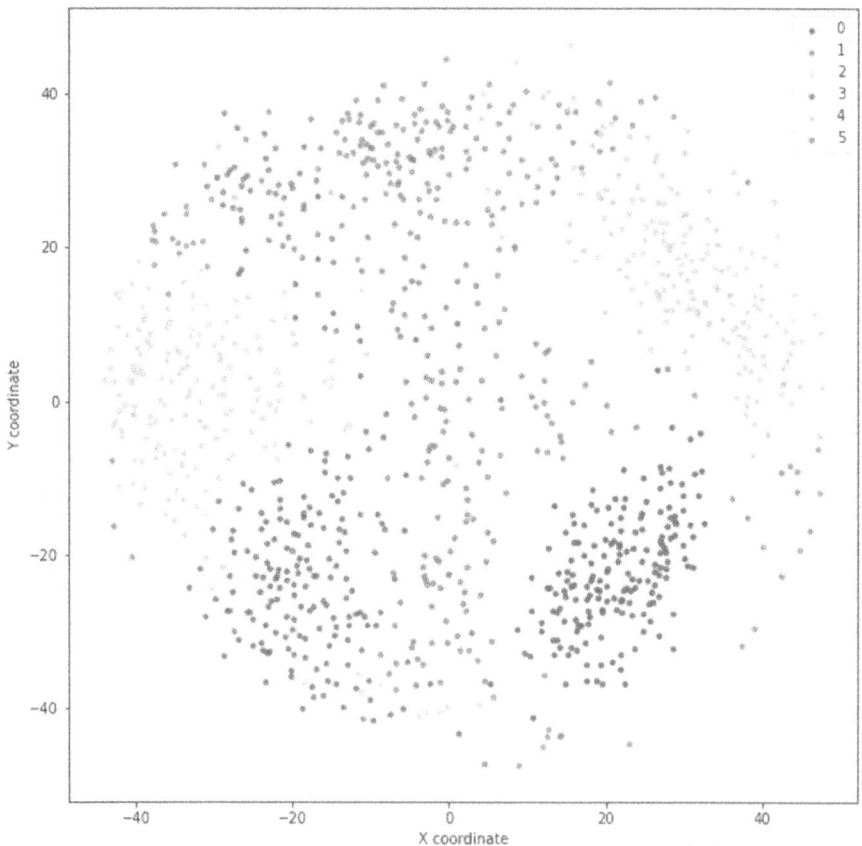

FIGURE 6.5   Multidimensional Scaling on digits dataset.

REFERENCES

1. Steyvers, M. (2006). Multidimensional scaling. *Encyclopedia of cognitive science*, Wiley, USA.
2. Bengio, Y., Paiement, J. F., Vincent, P., Delalleau, O., Roux, N. L., & Ouimet, M. (2004). Out-of-sample extensions for LLE, isomap, mds, eigenmaps, and spectral clustering. In *Proceedings of advances in neural information processing systems* (pp. 177–184). Vancouver, British Columbia, Canada.
3. Saul, L. K., Weinberger, K. Q., Sha, F., Ham, J., & Lee, D. D. (2006). Spectral methods for dimensionality reduction. *Semi-supervised learning, 3*, pp. 293–308.
4. Burges, C. J. (2009). Geometric methods for feature extraction and dimensional reduction—a guided tour. In *Data mining and knowledge discovery handbook* (pp. 53–82). Springer, Boston, MA.

7 Isomap

7.1 EXPLANATION AND WORKING

Let us first understand the concept of geodesic distance since it is essential in nonlinear dimensionality reduction. As we already discussed in Chapter 6, Multidimensional Scaling (MDS) tries to maintain the Euclidean distance in the lower dimension. However, in some cases preserving Euclidean distance in the lower dimension while carrying out dimensionality reduction might not give us the desired result. The Euclidean metric for distance works only if the neighborhood structure can be approximated as a linear structure in nonlinear manifolds.

Let us say that the neighborhood structure consists of holes. In such cases, Euclidean distances will be highly misleading. Contrary to this, if we measure the distance between points by traversing the manifold, we can have a better approximation of how near or far any two points on the manifold are [1]. Let us understand this concept using a simplified example. Let us assume our data lies on a two-dimensional circular manifold structured as shown in Figure 7.1.

Why is the geodesic distance a better fit than the Euclidean distance in a nonlinear manifold?

As seen in Figure 7.1, the two-dimensional data is reduced to one dimension, using both Euclidean distances and approximate geodesic distances. Now, if we check the 1D mapping based on the Euclidean metric, we observe that the two very distant points (in this case, the points a and b) have been mapped poorly. As stated earlier, only the points that lie on the neighborhood structure can be approximated as a linear structure (in this case, the points c and d) to give satisfactory results. On the other hand, if we check the 1D mapping based on geodesic distances, we can observe that it rightly approximates the close points as neighbors and the far away points as distant.

The geodesic distances between two points can be approximated by graph distance between the two points, so, as we can see from the above discussion, even though the Euclidean metric does a relatively poor job in approximating the distance between two points in nonlinear manifolds, the geodesic metric of distances gives satisfactory results. Hence, while dealing with finding the approximated distance between points on a nonlinear manifold, using the geodesic metric of distances is a better fit. Isomap uses the concept of geodesic distance to solve the problem of dimensionality reduction.

Isomap stands for Isometric mapping. It is another distance-preserving nonlinear dimensionality reduction technique that is based on spectral theory. This technique solves nonlinear dimensionality reduction by preserving the geodesic distances between the points in the lower dimension [2].

Firstly, it finds the approximate distance between all the pairs of points on the higher dimensional manifold. Then it finds a set of y_t such that distances between

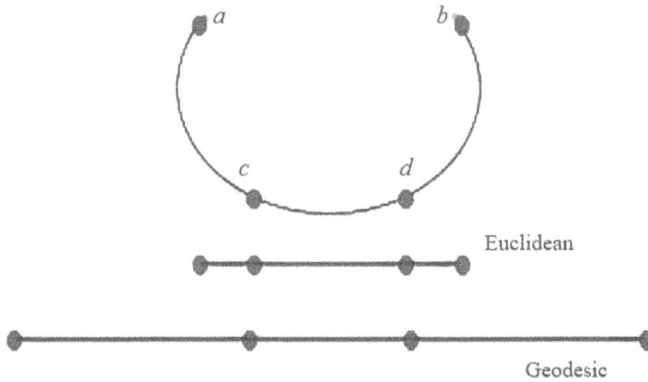

FIGURE 7.1 Representation of Euclidean and geodesic distances.

pairs of points in the low dimensional space are roughly the same as the distances between the pairs of points on the manifold [3]. Let us first see the steps involved in the algorithm and then later we discuss how these steps work

7.1.1 ISOMAP ALGORITHM

1. For every point in the original dataset, find it is k-Nearest Neighbor (with respect to the actual distance in the high-dimensional space).
2. Plot the k-nearest neighbor graph G = (V, E). Every point of the dataset is a vertex in the graph (hence, there n-vertices in total), and every point in this graph is connected to its k-nearest neighbors by an edge.
3. Compute the pairwise distances between all pairs of points in the graph using the graph's geodesic distance as the metric and represent it using a matrix (say A).
4. Find points in the lower-dimensional space such that pairwise distances between points are approximately the same as distances between the points on the graph.

The algorithm's output is the low dimensional representation of all the points computed in the final step.

The idea is that if we have enough points on the high dimensional manifold, that is if they are packed tightly, and because the manifold is locally Euclidean, the k-nearest neighbors on the original high dimensional space are also the manifold's k-nearest neighbors. We then form the nearest neighbor graph based on the idea that the shortest distances on the graph correspond to the manifold.

Now, how do we compute the pairwise shortest path between all the points on a graph? We either use the Floyd Warshall algorithm or Dijkstra's algorithm between all pairs to get the pairwise distance between all pairs of points on the manifold. Now we simply need to find points on low dimensional space such that Euclidean distances in this lower dimensional space are approximately equal to the shortest distance on the manifold between these points.

Hence, given all pairwise distances between n points, the question of how to find lower dimensional points whose distances are approximately equal to the given n^2 distances is precisely the problem solved by MDS [4].

7.2   ADVANTAGES AND LIMITATIONS

- The first step of Isomap depends on finding distances of the manifold. This step does not work accurately if we do not have dense enough points on the manifold. In such a case, the graph's distance could be very different from the distance obtained from the manifold.
- Another limitation of Isomap is that if the k for k-nearest Neighbor is too small, the graph may not even be connected, and if k is large, the graph might be too dense, leading us to find the wrong distances on the manifold. In general, finding the right k, even if it exists, is very hard.
- Isomap works very well when the data is strongly nonlinear.

7.3   USE CASES

- Isomap has a variety of practical applications. Medical applications of ultrasound imaging have expanded enormously over the last two decades. De-noising is a challenging issue for better medical interpretation and diagnosis on a high volume of data sets in echocardiography. The manifold learning algorithm is applied on two-dimensional echocardiography images to discover the relationship between the frames of consecutive cycles of the heart motion. In this approach, each image can be depicted by a point on a reconstructed two-dimensional manifold by the Isomap algorithm and similar points can be related to similar images according to the property of periodic heartbeat cycle stand together [1].
- Another practical application of Isomap is interpolating images between frames of a video. Isomap can be used to map every frame of the video to a low dimension feature space that extracts meaningful features from the sequence of images. To insert new images between any two frames, images are selected from the original dataset that are mapped in the same area as those two frames [3].

To understand this method better, let us consider an example where we perform dimensionality reduction on high dimensional datasets using MDS and visualize the results.

7.4   EXAMPLES AND TUTORIAL

Example 1

While most of the real-world datasets like images and text data are very high dimensional, we will use the MNIST handwritten digits dataset for simplicity. The MNIST dataset is a collection of grayscale images of handwritten single digits between 0 and 9 that contains 60,000 images of size 28 × 28 pixels. Thus, this dataset has 60,000 data samples with a dimensionality of 784. To demonstrate dimensionality reduction on this dataset, we use Isomap to reduce the data's dimensionality and project the data onto a low dimensional feature space. This example will map the data with 784 features to two-dimensional feature space and visualize the results.

Let us import the MNIST handwritten digits dataset from the tensorflow library. Next, we will use the *sklearn.manifold* module from the scikit-learn library for dimensionality reduction. Finally, after applying Isomap on the dataset, we will plot the results to visualize the low dimensional representation of the data using the matplotlib library.

The first step is to import all the necessary libraries.

```
import tensorflow as tf
from tensorflow.examples.tutorials.mnist import input_data
import sklearn
from sklearn.manifold import Isomap
import matplotlib.pyplot as plt
```

Next, let us load the MNIST dataset.

```
mnist = input_data.read_data_sets("MNIST_data/")
```

Each image is of size 28 × 28 pixels which is flattened into a vector of size 784. Hence, mnist.train.images is an n-dimensional array (tensor) whose shape is [55000, 784], whereas, the shape of mnist.train.labels is [55000, 10] since there are 10 class labels from 0 to 9.

```
X_train = mnist.train.images
Y_train = mnist.train.labels
n_samples, n_features = X_train.shape
print(n_features)
print(n_samples)
```

Output:

```
784
55000
```

We use the Isomap algorithm. For this we use *sklearn.manifold.Isomap*. The parameter *n_components* denotes the dimensionality of the target projection space and *n_neighbors* denotes the number of nearest neighbors in the neighborhood graph. We then fit and transform the training data.

```
isomap =Isomap(n_components=2, n_neighbors = 10)
isomap.fit(X_train)
X_isomap = isomap.fit_transform(X_train)
```

Hence, the dimensionality of the projection space is reduced from 784 to 2. That is, the shape of X_isomap is [55000, 2]

```
n_samples, n_features = X_isomap.shape
print(n_features)
print(n_samples)
```

Output:

```
2
55000
```

Now, let us plot the reduced data using matplotlib where each data point is represented using a different color corresponding to its label.

```
plt.figure(figsize=(10,10))
plt.scatter(X_isomap[y==0, 0], X_isomap[y==0, 1],
color='blue', alpha=0.5,label='0', s=9, lw=2)
plt.scatter(X_isomap[y==1, 0], X_isomap[y==1, 1],
color='purple', alpha=0.5,label='1',s=9, lw=2)
plt.scatter(X_isomap[y==2, 0], X_isomap[y==2, 1],
color='yellow', alpha=0.5,label='2',s=9, lw=2)
plt.scatter(X_isomap[y==3, 0], X_isomap[y==3, 1],
color='black', alpha=0.5,label='3',s=9, lw=2)
plt.scatter(X_isomap[y==4, 0], X_isomap[y==4, 1],
color='gray', alpha=0.5,label='4',s=9, lw=2)
plt.scatter(X_isomap[y==5, 0], X_isomap[y==5, 1],
color='turquoise', alpha=0.5,label='5',s=9, lw=2)
plt.scatter(X_isomap[y==6, 0], X_isomap[y==6, 1], color='red',
alpha=0.5,label='6',s=9, lw=2)
plt.scatter(X_isomap[y==7, 0], X_isomap[y==7, 1],
color='green', alpha=0.5,label='7',s=9, lw=2)
plt.scatter(X_isomap[y==8, 0], X_isomap[y==8, 1],
color='violet', alpha=0.5,label='8',s=9, lw=2)
plt.scatter(X_isomap[y==9, 0], X_isomap[y==9, 1],
color='orange', alpha=0.5,label='9',s=9, lw=2)
plt.ylabel('Y coordinate')
plt.xlabel('X coordinate')
plt.legend()
plt.show()
```

Figure 7.2 shows the visualization of the MNIST data reduced to a two-dimensional
feature space using Isomap.

Example 2

To demonstrate how Isomap works as a manifold learning technique, let us consider an illustration of dimensionality reduction on an S-curve dataset. This toy
dataset has 2000 data points lying on an S-shaped manifold. The objective of
manifold learning is to learn the underlying intrinsic geometry of the manifold and
unfold the manifold in low dimensional space by preserving the locality of the
data points.

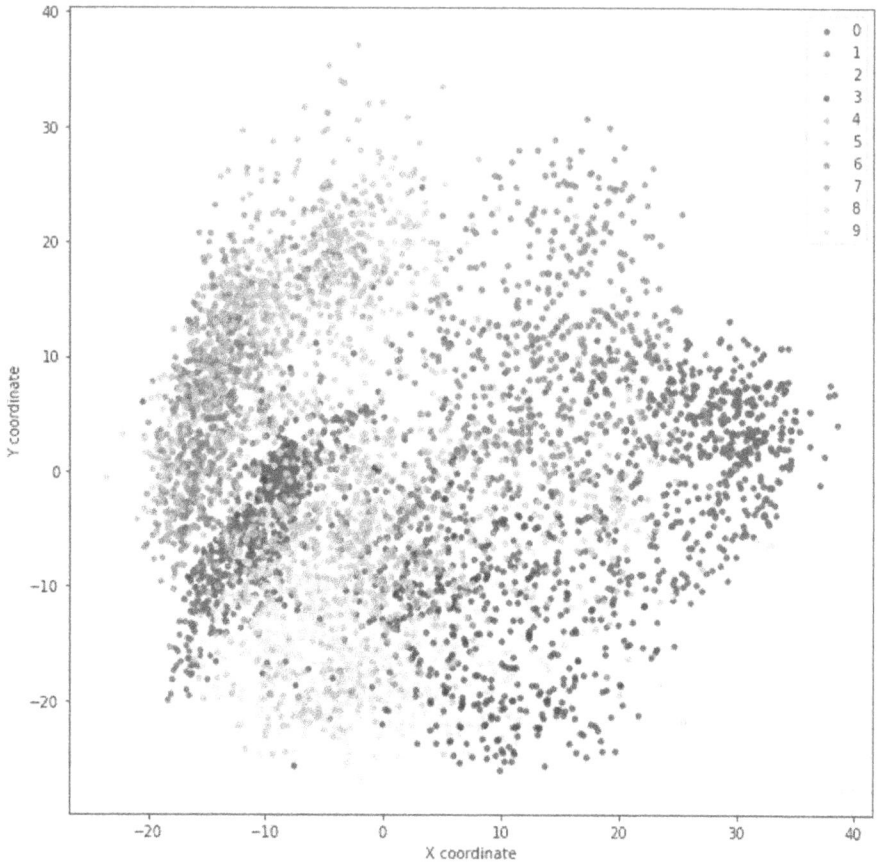

FIGURE 7.2   Isomap on MNIST data.

For this example, we use sklearn to create an S-curve dataset with 2000 data points. Likewise, we use *Isomap* algorithm implementation from the *sklearn,manifold* module. Finally, to plot the data, we use matplotlib.

Firstly, import all the required libraries.

```
from sklearn.manifold import Isomap
from sklearn import datasets
import matplotlib.pyplot as plt
from matplotlib.ticker import NullFormatter
```

Next, load the S-curve dataset with 2000 data points *(n_points)* using *datasets. make_s_curve.*

```
n_points = 2000
X, color = datasets.make_s_curve(n_points, random_state=0)
```

Now, create a 3D plot of the S-curve dataset.

```
fig = plt.figure(figsize=(45, 25))
ax = fig.add_subplot(251, projection='3d')
ax.scatter(X[:, 0], X[:, 1], X[:, 2], c=color, cmap=plt.
cm.jet, s=9, lw=1)
ax.view_init(10, -72)
```

Figure 7.3 is the visualization of the S-curve dataset on a three-dimensional scatter plot.

Next, we use Isomap with dimensionality of the target projection space *(n_components)* as 2 and number of neighbors *(n_neighbors)* as 10.

```
isomap= Isomap (n_components=2, n_neighbors = 10)
X_isomap = isomap.fit_transform(X)
```

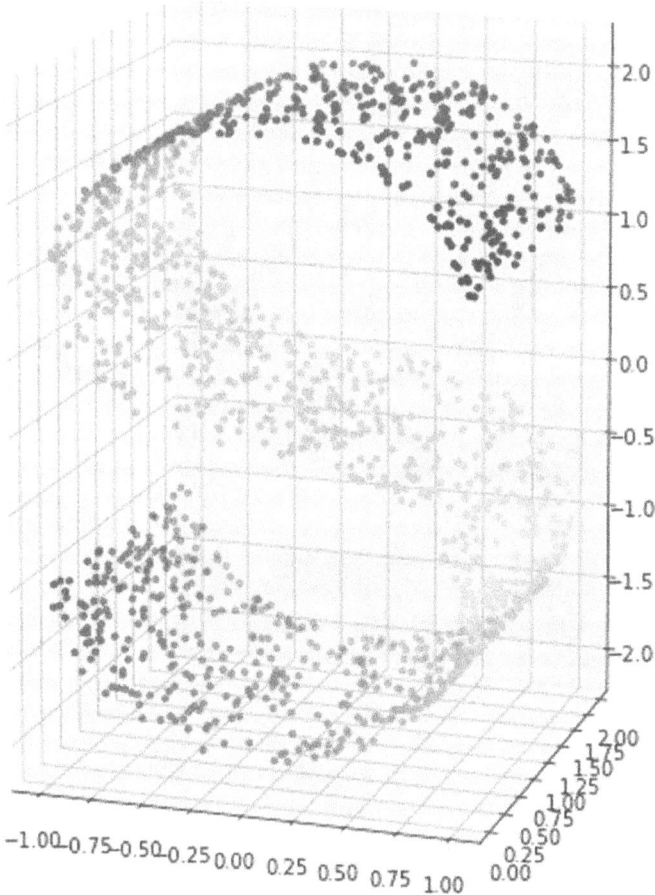

FIGURE 7.3 S-shaped curve, n=2000.

After fitting and transforming the data, plot the results.

```
fig = plt.figure(figsize=(5, 5))
ax = fig.add_subplot(1,1,1)
ax.scatter(X_isomap[:, 0], X_isomap[:, 1], c=color, cmap=plt.
cm.jet, s=9, lw=1)
ax.xaxis.set_major_formatter(NullFormatter())
ax.yaxis.set_major_formatter(NullFormatter())
ax.axis('tight')
plt.ylabel('Y coordinate')
plt.xlabel('X coordinate')
plt.show()
```

Figure 7.4 illustrates the result of manifold learning by ISOMAP on the S-curve dataset with 2000 points.

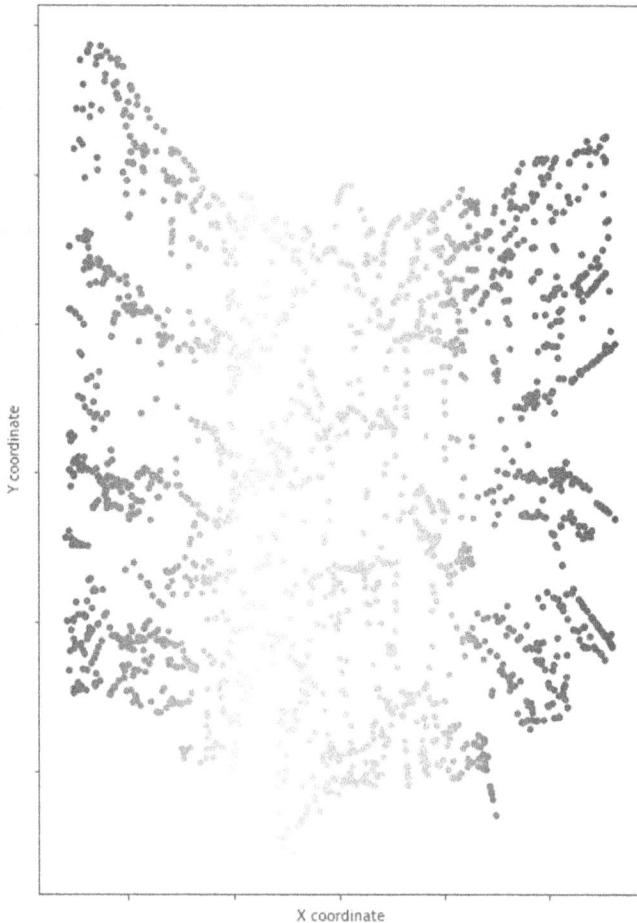

FIGURE 7.4   Isomap on S-curve data.

Example 3

For this example, we use sklearn to create a Swiss roll dataset with 2000 data points. Likewise, we use *Isomap* algorithm implementation from the *sklearn,manifold* module. Finally, to plot the data, we use matplotlib.
   Firstly, import all the required libraries.

```
from sklearn.manifold import Isomap
from sklearn import datasets
import matplotlib.pyplot as plt
from matplotlib.ticker import NullFormatter
```

Next, load the Swiss roll dataset with 2000 data points *(n_points)* using *datasets. make_swiss_roll.*

```
n_points = 2000
X, color = datasets.make_swiss_roll(n_points,
random_state=0)
```

Now, create a 3D plot of the Swiss roll dataset.

```
fig = plt.figure(figsize=(45, 25))
ax = fig.add_subplot(251, projection='3d')
ax.scatter(X[:, 0], X[:, 1], X[:, 2], c=color, cmap=plt.
cm.jet, s=9, lw=1)
ax.view_init(10, -72)
```

Figure 7.5 is the visualization of the Swiss roll dataset on a three-dimensional scatter plot.
   Next, we use Isomap with dimensionality of the target projection space *(n_components)* as 2 and number of neighbors *(n_neighbors)* as 10.

```
isomap = Isomap (n_components=2, n_neighbors = 10)
X_isomap = isomap.fit_transform(X)
```

After fitting and transforming the data, plot the results.

```
fig = plt.figure(figsize=(5, 5))
ax = fig.add_subplot(1,1,1)
ax.scatter(X_isomap[:, 0], X_isomap[:, 1], c=color, cmap=plt.
cm.jet, s=9, lw=1)
ax.xaxis.set_major_formatter(NullFormatter())
ax.yaxis.set_major_formatter(NullFormatter())
ax.axis('tight')
plt.ylabel('Y coordinate')
plt.xlabel('X coordinate')
plt.show()
```

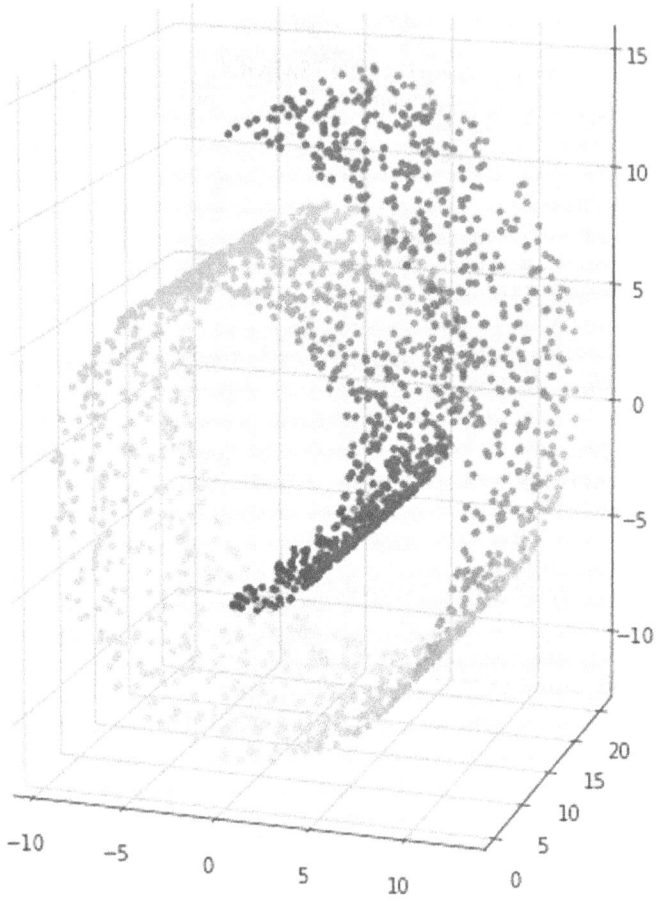

FIGURE 7.5   Swiss roll dataset, n=2000.

Figure 7.6 illustrates the result of manifold learning by Isomap on the Swiss roll dataset with 2000 points.

Example 4

Next, let us see the results of dimensionality reduction on the 64-dimensional handwritten digits dataset (not to be confused with the MNIST handwritten digits database) consisting of 1083 images of six handwritten digits (0–5) each of size 8 × 8 pixels.

Let us import the digits dataset from sklearn library. Similar to the previous example, we will use the sklearn.manifold module from the Scikit-learn library for dimensionality reduction and matplotlib for plotting the data.

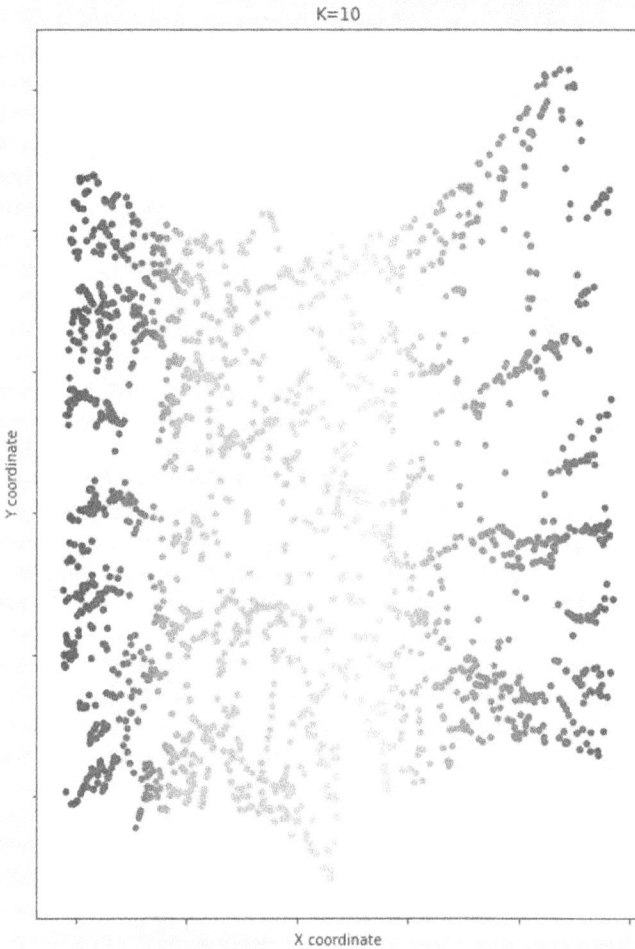

FIGURE 7.6 Isomap on Swiss roll dataset.

First, import all the necessary libraries.

```
import sklearn
from sklearn import datasets
from sklearn.manifold import Isomap
import matplotlib.pyplot as plt
```

Next, load the digits dataset using *load_digits*. This dataset originally contains 1797 data samples. That is, it contains 1797 images of digits. Each data point is an image of a digit of size 8×8. Hence, the dimensionality of the data is 64. The dataset is divided into 10 classes (digits from 0 to 9). Thus, the shape of the data is [1797, 64].

However, in this example we will consider only 6 classes (digits with label 0–5) which contain 1083 samples.

```
digits = datasets.load_digits(n_class=6)
X = digits.data
y = digits.target
n_samples, n_features = X.shape
print(n_features)
print(n_samples)
```

Output:

```
64
1083
```

Here, digits.data contains the images and digits.target contains the corresponding labels similar to mnist.train.images and mnist.train.labels, respectively.

We use *sklearn.manifold.Isomap* for Isomap. The parameter *n_components* denotes the dimensionality of the target projection space and the parameter *n_neighbors* denotes the number of nearest neighbors in the neighborhood graph. We then fit and transform the training data.

```
isomap = Isomap(n_components=2, n_neighbors = 10)
X_isomap = isomap.fit_transform(X)
```

Now, plot the transformed data using matplotlib.

```
plt.figure(figsize=(10,10))
plt.scatter(X_isomap[y==0, 0], X_isomap[y==0, 1],
color='blue', alpha=0.5,label='0', s=9, lw=2)
plt.scatter(X_isomap[y==1, 0], X_isomap[y==1, 1],
color='green', alpha=0.5,label='1',s=9, lw=2)
plt.scatter(X_isomap[y==2, 0], X_isomap[y==2, 1],
color='orange', alpha=0.5,label='2',s=9, lw=2)
plt.scatter(X_isomap[y==3, 0], X_isomap[y==3, 1],
color='purple', alpha=0.5,label='3',s=9, lw=2)
plt.scatter(X_isomap[y==4, 0], X_isomap[y==4, 1],
color='violet', alpha=0.5,label='4',s=9, lw=2)
plt.scatter(X_isomap[y==5, 0], X_isomap[y==5, 1], color='red',
alpha=0.5,label='5',s=9, lw=2)
plt.ylabel('Y coordinate')
plt.xlabel('X coordinate')
plt.legend()
plt.show()
```

Figure 7.7 visualizes the two-dimensional embedding obtained after performing dimensionality reduction on the 64-dimensional data consisting of 1083 data samples using Isomap.

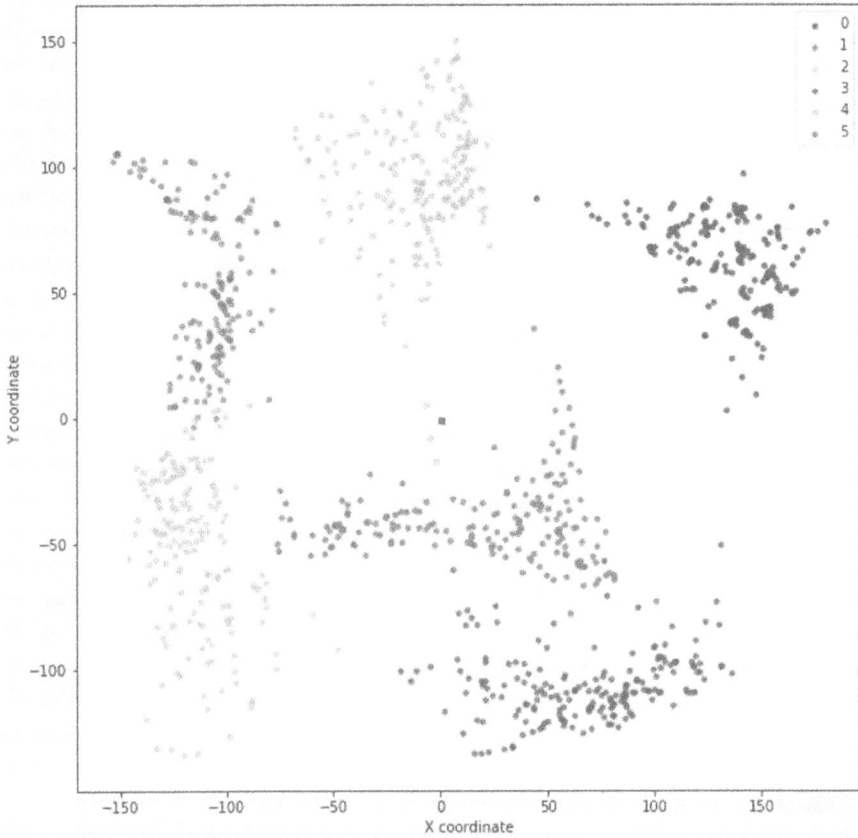

FIGURE 7.7 Isomap on digits dataset.

REFERENCES

1. Gifani, P., Behnam, H., Shalbaf, A., & Sani, Z. A. (2011, February). Noise reduction of echocardiography images using Isomap algorithm. In *2011 1st Middle East conference on biomedical engineering* (pp. 150–153). IEEE, Sharjah, UAE.
2. Burges, C. J. (2009). Geometric methods for feature extraction and dimensional reduction—a guided tour. In *Data mining and knowledge discovery handbook* (pp. 53–82). Springer, Boston, MA.
3. Robaszkiewicz, S., & El Ghazzal, S. (2013). Interpolating images between video frames using non-linear dimensionality reduction. In *Proceedings of the 30th international conference on machine learning*, Atlanta, GA.
4. Bengio, Y., Paiement, J. F., Vincent, P., Delalleau, O., Roux, N. L., & Ouimet, M. (2004). Out-of-sample extensions for LLE, isomap, MDS, eigenmaps, and spectral clustering. In *proceedings of advances in neural information processing systems* (pp. 177–184). Vancouver, British Columbia, Canada.

8 Random Projections

8.1 EXPLANATION AND WORKING

In application areas such as computer vision and pattern recognition where the dataset not only has a large number of data points but also has data of very high dimensionality, PCA might not be a feasible technique as it becomes computationally expensive to handle the entire data matrix [1]. In such cases, random projection proves to be a powerful method for dimensionality reduction that creates compact representations of high dimensional data, preserving well the distances between data points. This method involves choosing a random subspace for projection that is independent of the input data by using a projection matrix with its entries being randomly sampled and at the same time exhibits substantial computational efficiency and accuracy in projecting data from a very high dimension to a lower dimensional space when compared to other dimensionality reduction methods like PCA. Random projections deal with high dimensional data by mapping them into a lower dimensional space while they guarantee approximate preservation of distances between data points in the lower dimensional space.

Let $X \in \mathbb{R}^d$ be an $n \times d$ matrix of n data points in high dimensional space d. We choose a randomly sampled $d \times k$ projection matrix, W, and define the projection of X in lower dimensional space k to be

$$Y = XW \tag{8.1}$$

where $Y \in \mathbb{R}^k$ is an $n \times k$ matrix that gives the k-dimensional approximations of the n data points.

Here W is a $d \times k$ matrix with entries w_{ij} sampled independently at random using distributions such as the Gaussian distribution. The projection matrix W can also be sampled from various other distributions as follows:

$$w_{ij} = \begin{cases} +1, & p = 1/2 \\ -1, & p = 1/2 \end{cases} \tag{8.2}$$

and

$$w_{ij} = \sqrt{3} \begin{cases} +1, & p = 1/6 \\ 0, & p = 2/3 \\ -1, & p = 1/6 \end{cases} \tag{8.3}$$

While the random projection matrix can also be drawn from a Gaussian distribution, the above distribution called the Achliopta's distribution [2] is much easier to compute, requiring only a uniform random generator, as the matrix is two-thirds sparse, thus reducing the computation significantly.

With PCA, we find the projected data by computing the covariance matrix and decomposing it into its singular value form which is computation intensive for large data. However, unlike PCA, random projection is fairly simple to compute as it just involves the multiplication of a matrix that can be generated using a simple procedure when compared to the computationally intensive steps involved in PCA. One important property of random projection is that it most approximately preserves the lengths of the original data points as well as the pairwise distances between the original points when mapped to a lower dimensional space up to an arbitrary approximation factor. The Johnson-Lindenstrauss lemma [3] that deals with Euclidean distance-preserving embeddings serves as the basis for projections.

In PCA, the goal is to capture the direction along which the variance is maximized and project the data onto those directions, whereas random projections do not take into account the original data at all. Instead, a random subspace is chosen for projection though this seems counter-intuitive. The optimality of PCA's reduction comes at a significant cost – the high computational complexity associated with eigenanalysis in the case where the dimensionality of the data as well as the number of data points is very high [1]. In PCA, the local distances are not given importance. That is, the pairwise distances between points may be arbitrarily distorted as PCA's objective is to minimize the overall distortion. Hence, there is no assurance that the distance between any two points in the original space will be precisely the same as that in the projected lower dimensional space. However, random projection guarantees to preserve pairwise distances between points in the reduced space with a high probability.

Comparison between PCA and random projections:

a. Accuracy: Though PCA tends to be the most optimal and accurate dimensionality reduction technique, random projections give results that are comparable to that of PCA [1, 4].
b. Computational complexity: As the dimensions of the data increases, performing PCA becomes computationally expensive, On the other hand, the computational complexity of random projections is significantly lower even while handling data with very high dimensionality.
c. Distance preservation: PCA aims at minimizing the overall distortion and not the pairwise distances between the points. Whereas, random projection provides guarantee that distance between any two points is preserved when projected in a lower dimensional space.

One interesting thing about random projection is that it gives a fairly accurate bound on the lowest dimension we can reduce the data to while still preserving the distance up to a factor of ε. In other dimensionality reduction techniques, there is no way other than experimentation using trial and error to determine the lowest dimension to which we can project down, while not incurring too much error. But the Johnson-Lindenstrauss lemma formalizes the lowest dimension guarantee of the random projection method.

The elementary proof of the Johnson-Lindenstrauss lemma provided by [5], which is a simplification of the original proof of the Johnson-Lindenstrauss lemma, states that for any $0 < \varepsilon < 1$ and any integer n let k be a positive integer such that

$$k \geq \frac{4}{\frac{\varepsilon^2}{2} - \frac{\varepsilon^3}{3}} \log n \tag{8.4}$$

Then for any set X of n points $\in \mathbb{R}^d$ there exists a mapping $f : \mathbb{R}^d \to \mathbb{R}^k$ such that for all $x_i, x_j \in X$:

$$(1 - \varepsilon) \left\| x_i - x_j \right\|^2 \leq \left\| f(x_i) - f(x_j) \right\|^2 \leq (1 + \varepsilon) \left\| x_i - x_j \right\|^2 \tag{8.5}$$

This lemma guarantees that when any n points in a high dimensional Euclidean space d are mapped onto k dimensions where k is at least $\frac{4}{\frac{\varepsilon^2}{2} - \frac{\varepsilon^3}{3}} \log n$, the Euclidean distance between any two points is preserved without distorting it by more than a factor of $1 \pm \varepsilon$. This is surprising as the original dimension d is not taken into consideration while choosing k, and it is only the number of data points n that matter. Hence, the same k holds for same choice of n irrespective of what the original dimension is.

Next, let us discuss some of the fundamental properties of projections [6] on length preservation and distance preservation.
Projection Lemmas:

$$w_{ij} = \sqrt{3} \begin{cases} +1, & p = 1/6 \\ 0, & p = 2/3 \\ -1, & p = 1/6 \end{cases} \tag{8.6}$$

The above random distribution has zero mean and unit variance. That is,

$$E(w_{ij}) = 0, \tag{8.7}$$

$$Var(w_{ij}) = 1 \tag{8.8}$$

Let the input $X \in \mathbb{R}^d$ be an $n \times d$ matrix:

$$X = \begin{bmatrix} x_1 \\ x_2 \\ \vdots \\ x_n \end{bmatrix}$$

where x_i represents the d-dimensional data points.

Let $W \in \mathbb{R}^k$ be the $d \times k$ projection matrix:

$$W = \begin{bmatrix} w_1 & w_2 & \cdots & w_k \end{bmatrix}$$

where $w_i's$ are the entries of the projection matrix that are randomly sampled from the distribution. Then the projection of X on a k dimensional space is given by:

$$Y = \frac{1}{\sqrt{k}} A.R \tag{8.9}$$

$$= \begin{bmatrix} x_1 w_1^T & x_1 w_2^T & \cdots & x_1 w_k^T \\ x_2 w_1^T & x_2 w_2^T & \cdots & x_2 w_k^T \\ \vdots & & \vdots & \\ x_n w_1^T & x_n w_2^T & \cdots & x_n w_k^T \end{bmatrix} \tag{8.10}$$

$$= \begin{bmatrix} y_1 \\ y_2 \\ \vdots \\ y_n \end{bmatrix} \tag{8.11}$$

Thus, the projection of the ith point is represented as:

$$y_1 = \frac{1}{\sqrt{k}} \begin{bmatrix} x_i w_1^T & x_i w_2^T & \cdots & x_i w_k^T \end{bmatrix} \tag{8.12}$$

Lemma 8.1: Length Preservation

Let the entries of W sampled randomly from the distribution have zero mean and unit variance. Then,

$$E\left(\|y_i\|^2\right) = \|x_i\|^2 \tag{8.13}$$

That is, the length of the original vectors is preserved when projected onto a lower dimensional space.

Proof:
We know by (8.7) and (8.8) that,

$$E(w_{ij}) = 0 \text{ and}$$

$$Var(w_{ij}) = 1$$

For $i = 1$,

$$\|y_i\|^2 = \frac{1}{k} \sum_i (x_1 w_i^T)^2$$

$$= \frac{1}{k} \sum_i \alpha_i^2$$

where $\alpha_i = x_1 w_i^T = \sum_j x_{1j} w_{ji}$. Therefore,

$$E(\|y_i\|^2) = E\left(\frac{1}{k} \sum_i \alpha_i^2 \right)$$

$$= \frac{1}{k} \sum_i E(\alpha_i^2)$$

$$= \frac{1}{k} \sum_i var(\alpha_i) + (E(\alpha_i))^2$$

$$= \frac{1}{k} \sum_i var\left(\sum_j x_{1j} w_{ji} \right) + \left(E\left(\sum_j x_{1j} w_{ji} \right) \right)^2$$

$$= \frac{1}{k} \sum_i \left(\left(\sum_j var(x_{1j} w_{ji}) \right) + \left(\sum_j E(x_{1j} w_{ji}) \right)^2 \right)$$

[since variance satisfies linearity in the case of independent variables. That is,

$var\left(\sum_i X_i \right) = \sum_i Var(X_i)$ if $X_i's$ are independent. Similarly, expectation also fol-

lows linearity, $E\left(\sum_i X_i \right) = \sum_i E(X_i)$]

[Only X_i s will do]

$$= \frac{1}{k} \sum_i \left(\left(\sum_j x_{1j}^2 Var\left(w_{ji}\right) \right) + \left(\sum_j x_{1j} E\left(w_{ji}\right) \right)^2 \right)$$

[since $var(aX+b) = a^2 Var(X)$ for any $a,b \in \mathbb{R}$. Also, $E(aX) = aE(X)$]

$$= \frac{1}{k} \sum_i \left(\sum_j x_{1j}^2 + 0 \right)$$

[By (8.7) and (8.8)]

$$= \frac{1}{k} \sum_i \|x_1\|^2 = \|x_1\|^2$$

Therefore,

$$E\left(\|y_i\|^2\right) = \|x_i\|^2$$

That is, for any choice of projection matrix W whose mean of its entries is zero and variance is one, the length of the original vectors is preserved in the lower dimensional projected space.

Lemma 8.2: Distance Preservation

The results on length preservation can be generalized to distance preservation. Let the entries of projection matrix W have zero mean and unit variance. Then,

$$E\left(\|y_i - y_j\|^2\right) = \|x_i - x_j\|^2 \tag{8.14}$$

Proof:
Using the result of the previous lemma, we can deduce this:

$$y_1 - y_2 = \frac{1}{\sqrt{k}} \left[(x_1 - x_2) w_1^T \ (x_1 - x_2) w_2^T \cdots (x_1 - x_2) w_k^T \right]$$

$$= \frac{1}{\sqrt{k}} \left[uw_1^T \ uw_2^T \cdots uw_k^T \right]$$

where $u = (x_1 - x_2)$.

$$\|y_i - y_j\|^2 = \frac{1}{k}\sum_i (uw_i^T)^2$$

$$= \frac{1}{k}\sum_i \alpha_i^2$$

where $\alpha_i = uw_i^T = \sum_j u_j w_{ji}$.

$$E\left(\|y_i - y_j\|^2\right) = E\left(\frac{1}{k}\sum_i \alpha_i^2\right)$$

$$= \frac{1}{k}\sum_i E(\alpha_i^2)$$

$$= \frac{1}{k}\sum_i var(\alpha_i) + \left(E(\alpha_i)\right)^2$$

$$= \frac{1}{k}\sum_i var\left(\sum_j u_j w_{ji}\right) + \left(E\left(\sum_j u_j w_{ji}\right)\right)^2$$

$$= \frac{1}{k}\sum_i \left(\left(\sum_j var(u_j w_{ji})\right) + \left(\sum_j E(u_j w_{ji})\right)^2\right)$$

$$= \frac{1}{k}\sum_i \left(\left(\sum_j u_j^2 Var(w_{ji})\right) + \left(\sum_j u_j E(w_{ji})\right)^2\right)$$

$$= \frac{1}{k}\sum_i \left(\sum_j u_j^2 + 0\right)$$

$$= \frac{1}{k}\sum_i \|u\|^2$$

$$= \|u\|^2 = \|(x_1 - x_2)\|^2$$

Therefore,

$$E\left(\|y_i - y_j\|^2\right) = \|x_i - x_j\|^2$$

That is, the distance between any two points in the high dimensional space d is preserved when projected onto the lower dimensional space k. Hence, random projection guarantees pairwise distance preservation in the projected low dimensional space.

8.2 ADVANTAGES AND LIMITATIONS

- Firstly, unlike the traditional methods, random projections do not suffer from the curse of dimensionality.
- As discussed earlier in Section 8.1, when the dimension of the data is very high, random projections are computationally efficient and have less time complexity as compared to PCA which involves computation-intensive steps. It is much simpler to generate a random projection matrix and compute the projections that just involve simple matrix multiplication, unlike PCA where it becomes computationally expensive to find the covariance matrix and perform singular value decomposition as the dimensionality increases. This computational complexity associated with very high dimensional data is overcome by random projections. This reduction in the computation cost is the main advantage of random projections over PCA.
- On the other hand, another advantage of random projections is that they provide a guarantee for pairwise distance preservation in low dimensional space. Also, random projections can be used for clustering high dimensional data.
- But the drawback of this is that different random projections lead to different clustering outcomes, making this technique highly unstable for clustering.

8.3 USE CASES

Random projection is a technique that is widely used in various machine learning applications [7]. Random projection being a computationally efficient and accurate method for dimensionality reduction of very high dimensional data makes it a good choice for dimensionality reduction on high dimensional image and text data [1]. This method of dimensionality reduction is also widely used in various computer vision problems [8]. Specifically, random projection is used as a dimensionality reduction tool in face recognition [9]. The application areas of random projection also include information retrieval from text documents and processing of noisy and noiseless images. Another realistic application of this technique is in data mining problems involving a large amount of high dimensional data [1].

8.4 EXAMPLES AND TUTORIAL

To understand this algorithm better, let us consider a few simple examples where we perform dimensionality reduction on high dimensional datasets using random projection and visualize the results.

Example 1

While most of the real-world datasets like images and text data are very high dimensional in nature, we will use the MNIST handwritten digits dataset for the sake of simplicity. The MNIST dataset is a collection of grayscale images of handwritten single digits between 0 and 9 that contains 55,000 images of size

28 × 28 pixels. Thus, this dataset has 55,000 data samples with a dimensionality of 784. To demonstrate dimensionality reduction on this dataset, we use random projection to reduce the dimensionality of the data and project the data onto a low dimensional feature space. In this example, we will map the data with 784 features to a two-dimensional feature space and visualize the results.

Let us import the MNIST handwritten digits dataset from the tensorflow library. Next, we will use the *sklearn.random_projection* module from the scikit-learn library for dimensionality reduction. Finally, after applying random projection on the dataset, we will plot the results to visualize the low dimensional representation of the data using the matplotlib library.

The first step is to import all the necessary libraries.

```
import tensorflow as tf
from tensorflow.examples.tutorials.mnist import input_data
import sklearn
from sklearn.random_projection import GaussianRandomProjection
from sklearn.random_projection import SparseRandomProjection
import matplotlib.pyplot as plt
```

Next, let us load the MNIST dataset.

```
mnist = input_data.read_data_sets("MNIST_data/")
```

This dataset is split into three parts:

```
1. mnist.train - which has 55000 data points
2. mnist.test - which has 10000 data points
3. mnist.validation - which has 5000 data points
```

Every MNIST data point has two parts:

1. The handwritten digit image (X)
2. The corresponding class label (Y)

mnist.train.images contains all the training images and mnist.train.labels has all the corresponding training labels. As told earlier, each image is of size 28 × 28 pixels which is flattened into a vector of size 784. Hence, mnist.train.images is an n-dimensional array (tensor) whose shape is [55000, 784], whereas, the shape of mnist.train. labels is [55000, 10] since there are 10 class labels from 0 to 9.

```
X_train = mnist.train.images
Y_train = mnist.train.labels
n_samples, n_features = X_train.shape
print(n_features)
print(n_samples)
```

Output:

```
784
55000
```

First, we use the Gaussian random projection that projects the input space using a random matrix sampled from the Gaussian distribution $N(0,1)$. For this we use *sklearn.random_projection.GaussianRandomProjection*. The parameter *n_components* denotes the dimensionality of the target projection space. We then fit and transform the training data.

```
GaussianRP = GaussianRandomProjection(n_components=2,
random_state=2019)
GaussianRP.fit(X_train)
X_grp = GaussianRP.transform(X_train)
```

Hence, the dimensionality of the projection space is reduced from 784 to 2. That is, the shape of X_grp is [55000, 2]

```
n_samples, n_features = X_grp.shape
print(n_features)
print(n_samples)
```

Output:

```
2
55000
```

Now, let us plot the reduced data using matplotlib where each data point is represented using a different color corresponding to its label.

```
plt.figure(figsize=(10,10))
plt.scatter(X_grp[y_train==0, 0], X_grp[y_train==0, 1],
color='blue', alpha=0.5,label='0', s=9, lw=2)
plt.scatter(X_grp[y_train==1, 0], X_grp[y_train==1, 1],
color='purple', alpha=0.5,label='1',s=9, lw=2)
plt.scatter(X_grp[y_train==2, 0], X_grp[y_train==2, 1],
color='yellow', alpha=0.5,label='2',s=9, lw=2)
plt.scatter(X_grp[y_train==3, 0], X_grp[y_train==3, 1],
color='black', alpha=0.5,label='3',s=9, lw=2)
plt.scatter(X_grp[y_train==4, 0], X_grp[y_train==4, 1],
color='gray', alpha=0.5,label='4',s=9, lw=2)
plt.scatter(X_grp[y_train==5, 0], X_grp[y_train==5, 1],
color='lightgreen', alpha=0.5,label='5',s=9, lw=2)
plt.scatter(X_grp[y_train==6, 0], X_grp[y_train==6, 1],
color='red', alpha=0.5,label='6',s=9, lw=2)
plt.scatter(X_grp[y_train==7, 0], X_grp[y_train==7, 1],
color='green', alpha=0.5,label='7',s=9, lw=2)
plt.scatter(X_grp[y_train==8, 0], X_grp[y_train==8, 1],
color='lightblue', alpha=0.5,label='8',s=9, lw=2)
plt.scatter(X_grp[y_train==9, 0], X_grp[y_train==9, 1],
color='orange', alpha=0.5,label='9',s=9, lw=2)
plt.ylabel('Y coordinate')
plt.xlabel('X coordinate')
```

```
plt.legend()
plt.show()
```

Figure 8.1 shows the visualization of the MNIST data projected on a two-dimensional feature space using Gaussian random projection.

Next, we use the sparse random projection that uses a sparse random projection matrix whose entries are randomly drawn from Achliopta's distribution which is much more computationally efficient. For this we use *sklearn.random_projection. SparseRandomProjection.* The parameter *n_components* denotes the dimensionality of the target projection space. We then fit and transform the training data.

```
SparseRP = SparseRandomProjection(n_components=2,
random_state=2019)
SparseRP.fit(X_train)
X_srp = SparseRP.transform(X_train)
```

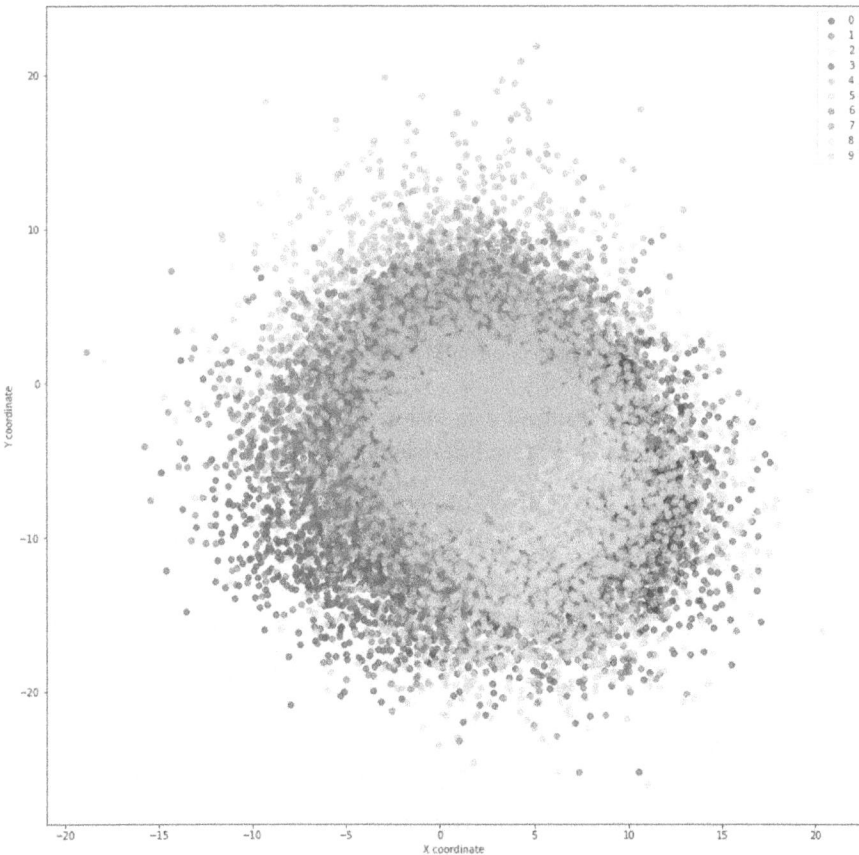

FIGURE 8.1 Gaussian random projection on MNIST data.

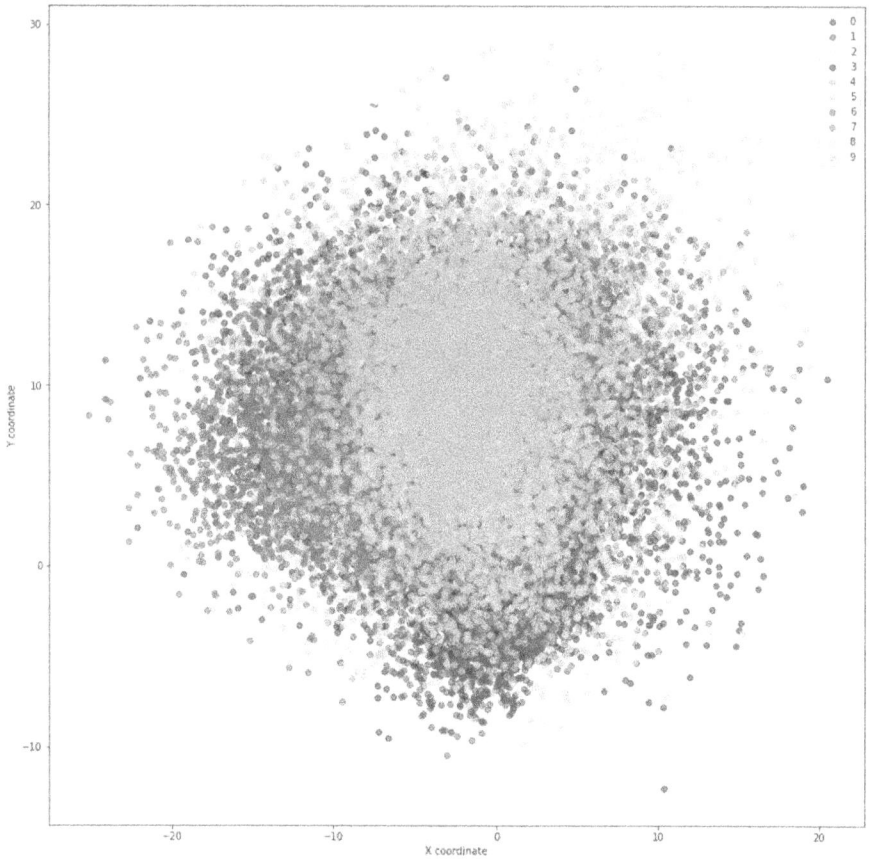

FIGURE 8.2   Sparse random projection on MNIST data.

Finally, let us plot the reduced data just as the data was plotted after applying Gaussian random projection. Figure 8.2 visualizes the result of sparse random projection on the MNIST dataset.

Example 2

Next, let us see the results of dimensionality reduction on the 64-dimensional handwritten digits dataset (not to be confused with the MNIST handwritten digits database) consisting of 1083 images of six handwritten digits (0–5) each of size 8 × 8 pixels.

Let us import the digits dataset from the sklearn library. Similar to the previous example, we will use the sklearn.random_projection module from the Scikit-learn library for dimensionality reduction and matplotlib for plotting the data.

First, import all the necessary libraries.

```
import sklearn
from sklearn import datasets
```

```
from sklearn.random_projection import GaussianRandomProjection
from sklearn.random_projection import SparseRandomProjection
import matplotlib.pyplot as plt
```

Next, load the digits dataset using *load_digits*. This dataset originally contains 1797 data samples. That is, it contains 1797 images of digits. Each data point is an image of a digit of size 8 × 8. Hence, the dimensionality of the data is 64. The dataset is divided into 10 classes (digits from 0 to 9). Thus, the shape of the data is [1797, 64]. However, in this example we will consider only 6 classes (digits with label 0–5) which contain 1083 samples.

```
digits = datasets.load_digits(n_class=6)
X = digits.data
y = digits.target
n_samples, n_features = X.shape
print(n_features)
print(n_samples)
```

Output:

```
64
1083
```

Here, digits.data contains the images and digits.target contains the corresponding labels similar to mnist.train.images and mnist.train.labels, respectively (as in the previous example).

We use *sklearn.random_projection.GaussianRandomProjection* for Gaussian random projection. The parameter *n_components* denotes the dimensionality of the target projection space. We then fit and transform the training data.

```
GaussianRP = GaussianRandomProjection(n_components=2,
random_state=2019)
X_grp = GaussianRP.fit_transform(X)
```

Now, plot the transformed data using matplotlib.

```
plt.figure(figsize=(10,10))
plt.scatter(X_grp[y==0, 0], X_grp[y==0, 1], color='blue',
alpha=0.5,label='0', s=9, lw=1)
plt.scatter(X_grp[y==1, 0], X_grp[y==1, 1], color='green',
alpha=0.5,label='1',s=9, lw=1)
plt.scatter(X_grp[y==2, 0], X_grp[y==2, 1], color='orange',
alpha=0.5,label='2',s=9, lw=1)
plt.scatter(X_grp[y==3, 0], X_grp[y==3, 1], color='purple',
alpha=0.5,label='3',s=9, lw=1)
plt.scatter(X_grp[y==4, 0], X_grp[y==4, 1], color='violet',
alpha=0.5,label='4',s=9, lw=1)
plt.scatter(X_grp[y==5, 0], X_grp[y==5, 1], color='red',
alpha=0.5,label='5',s=9, lw=1)
```

```
plt.ylabel('Y coordinate')
plt.xlabel('X coordinate')
plt.legend()
plt.show()
```

Figure 8.3 visualizes the two-dimensional embedding obtained after performing dimensionality reduction on the 64-dimensional data consisting of 1083 data samples using Gaussian random projections.

Next, use *sklearn.random_projection.SparseRandomProjection* for sparse random projection. Then, fit and transform the training data.

```
SparseRP = SparseRandomProjection(n_components=2,
random_state=2019)
X_srp = SparseRP.fit_transform(X)
```

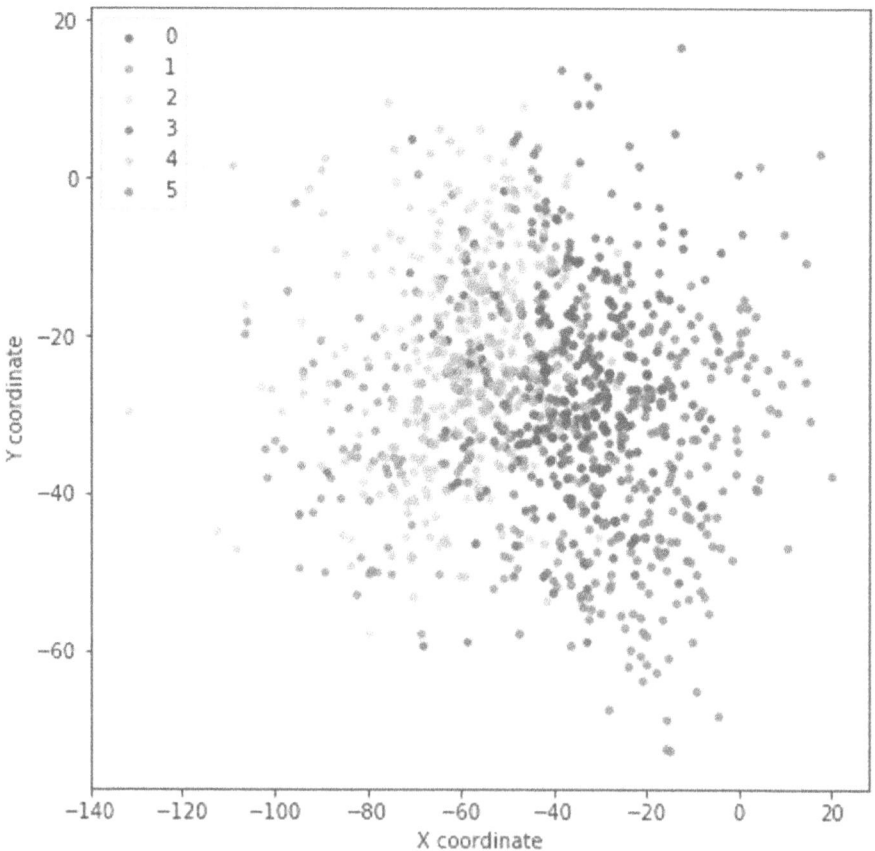

FIGURE 8.3   Gaussian random projection on digits dataset.

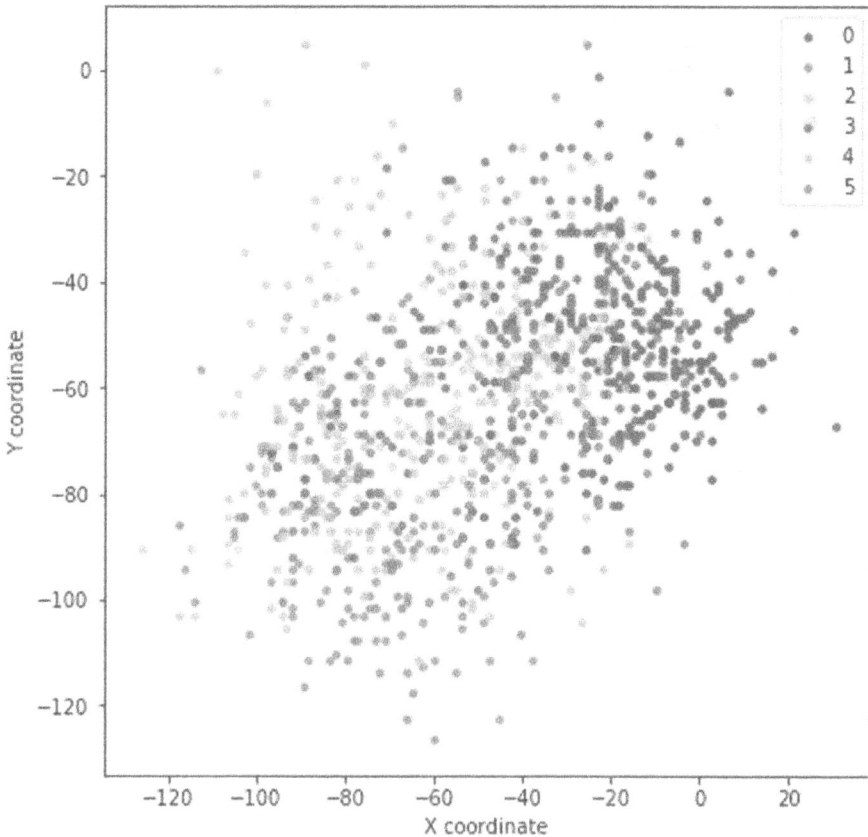

FIGURE 8.4   Sparse random projection on digits dataset.

Now, plot the transformed data using matplotlib similar to the Gaussian random projection case. Figure 8.4 is the visualization of the result of dimensionality reduction of the data using a sparse projection matrix randomly sampled from Achliopta's distribution.

REFERENCES

1. Bingham, E., & Mannila, H. (2001, August). Random projection in dimensionality reduction: applications to image and text data. In *Proceedings of the seventh ACM SIGKDD international conference on knowledge discovery and data mining* (pp. 245–250). San Francisco, CA.

2. Achlioptas, D. (2001, May). Database-friendly random projections. In *Proceedings of the twentieth ACM SIGMOD-SIGACT-SIGART symposium on principles of database systems* (pp. 274–281). Baltimore MD.

3. Johnson, W. B., & Lindenstrauss, J. (1984). Extensions of Lipschitz mappings into a Hilbert space. *Contemporary mathematics*, 26(1), 189–206.

4. Lin, J., & Gunopulos, D. (2003, May). Dimensionality reduction by random projection and latent semantic indexing. In *Proceedings of the text mining workshop at the 3rd SIAM international conference on data mining*, San Francisco, CA.

5. Dasgupta, S., & Gupta, A. (1999). An elementary proof of the Johnson-Lindenstrauss lemma. *International Computer Science Institute, technical report, 22*(1), 1–5.

6. Vempala, S. (2004). *The random projection method.* DIMACS Series in Discrete Mathematics and Theoretical Computer Science 65, AMS. ISBN 978-0-8218-3793-1.

7. Fradkin, D., & Madigan, D.(2003, August). Experiments with random projections for machine learning. In *Proceedings of the ninth ACM SIGKDD international conference on knowledge discovery and data mining* (pp. 517–522). Washington, DC.

8. Schu, G., Soldera, J., Medeiros, R., & Scharcanski, J. (2016). Random Projections and Their Applications in Computer Vision. DOI: 10.13140/RG.2.2.15444.40329.

9. Belghini, N., Zarghili, A., Kharroubi, J., & Majda, A. (2011, January). Sparse random projection and dimensionality reduction applied on face recognition. In *Proceedings of the international conference on intelligent systems & data processing* (pp. 78–82). Gujarat, India.

9 Locally Linear Embedding

9.1 EXPLANATION AND WORKING

In this section we will discuss an eigenvector-based unsupervised learning algorithm for the problem of nonlinear dimensionality reduction called Locally Linear Embedding (LLE) that computes low dimensional, neighborhood-preserving embeddings of high dimensional data [1]. It is an algorithm to find the low dimensional global coordinates of the data lying on a manifold embedded in a high dimensional space. LLE attempts to realize the nonlinear structure hidden in high dimensional data by taking advantage of the local symmetries of linear reconstructions. While projections by PCA and MDS do not preserve the neighborhood of data points, LLE very efficiently identifies the underlying structure of the manifold and ensures neighborhood preserving mapping of data in lower dimensional space [2].

Concisely, LLE has three major steps: identifying the neighborhood of the data points, finding the weights for reconstructing the data in that neighborhood, and lastly computing the coordinates best reconstructed by those weights in low dimension.

Let the data consist of n data points, each denoted by x_i of dimensionality d, lying on a smooth manifold, where each data point and its neighbors lie on a local linear patch of the manifold which is captured by LLE and projected onto a low dimensional space using neighborhood preserving mapping. In the original high dimensional space, the k-nearest neighbor graph for each data point x_i is identified and the following objective function that gives the summation of the squared distances between the data points and their reconstructions is minimized in order to find the optimal set of weights to reconstruct each data point in the reduced space from its k-nearest neighbors:

$$\varepsilon(x) = \sum_{i}^{n} \left| x_i - \sum_{j=1}^{k} w_{ij} x_{ij} \right|^2 \tag{9.1}$$

$$= \min_{w} \sum_{i}^{n} \left| x_i - \sum_{j=1}^{k} w_{ij} x_{ij} \right|^2 \tag{9.2}$$

Here, x_{ij} represents the jth neighbor of x_i and w_{ij} represents the weight (unknown) associated with the jth neighbor of x_i which is found by solving the objective function subject to the constraints that

1. $\sum_{j=1}^{k} w_{ij} = 1$ and

2. $w_{ij} = 0$ if data point x_j is not a k-nearest neighbor of x_i, that is, any data point x_i is reconstructed only from its k-nearest neighbors and the weights

associated with any other data point that doesn't belong to its k-nearest neighbor set are taken to be zero:

$$x_i = w_{i1}x_{i1} + w_{i2}x_{i2} + \cdots + w_{ik}x_{ik}$$

Once the weights are obtained, the points are reconstructed in the lower dimensional space from the nearest neighbors using these weights, thus preserving the local geometry of the locally linear patches of the underlying manifold in the low dimensional space as well. That is, after finding the weights for reconstruction, LLE constructs a neighborhood preserving mapping where the data points x_i in the higher dimensional space are mapped to lower dimensional observations y_i. This mapping in the low dimensional space is obtained by minimizing the following embedding cost function:

$$\varphi(y) = \sum_i^n \left| y_i - \sum_{j=1}^k w_{ij}y_j \right|^2 \tag{9.3}$$

$$= \min_y \sum_i^n \left| y_i - \sum_{j=1}^k w_{ij}y_j \right|^2 \tag{9.4}$$

where y_i represents the mapping of x_i in low dimensional space, which is to be found by optimizing the cost function. This is very similar to the previous cost function, but here the weights w_{ij} are fixed (obtained by optimizing the previous cost function) while we optimize y_i. Hence, the same set of weights can reconstruct data points in both high and low dimensional spaces as it is these weights around a data point that characterize the local geometry of the manifold [3].

To put it in a nutshell, in high dimensional space we reconstruct point x_i from its k-nearest neighbors, find the set of weights w_{ij} by minimizing the first cost function, and with those weights fixed, we go to the low dimensional space and find the configuration of the points in that space by optimizing the second cost function with fixed weights such that when the local patch is projected onto the low dimensional space, the neighborhood is preserved as we expect the same weights to reconstruct each of the data points from its neighbors; that is, each point y_j is reconstructed by its k-nearest neighbors with the same set of weights in the low dimensional space, thus preserving the local geometry in the reduced space as well.

LLE ALGORITHM

1. Find the k-nearest neighbors of each data point x_i in the original space.
2. Compute the set of weights w_{ij} that best reconstructs the data points x_i from their nearest neighbors by optimizing the first cost function.
3. Minimize the second cost function to compute each data point $x_i{'}s$ corresponding mapping in low dimensional space y_i, best reconstructed by the set of weights w_{ij} (obtained from step 2).

The optimization problem,

$$\min_{y} \sum_{i}^{n} \left| y_i - \sum_{j=1}^{k} w_{ij} y_j \right|^2$$

can be written in matrix form as

$$\min_{Y} |YI - YW|^2 \qquad (9.5)$$

where Y is a $p \times n$ matrix representing the n data points of dimensionality p (in low dimensional space), I is an $n \times n$ identity matrix, and W is an $n \times n$ matrix representing the weights where the ith column represents the weights associated with the neighbors of the data point x_i. Here, only the weights of its k-nearest neighbors have a value and other entries are zero:

$$\min_{Y} |YI - YW|^2$$

$$= \min_{Y} |Y(I - W)|^2$$

$$= \min_{Y} Tr\left((I - W)^T Y^T Y (I - W)\right)$$

[since Euclidean norm can be written in terms of trace as $|A|^2 = Tr\left(A^T A\right)$]

$$= \min_{Y} Tr\left(Y(I - W)(I - W)^T Y^T\right)$$

$$= \min_{Y} Tr\left(YMY^T\right) \qquad (9.6)$$

where $M = (I - W)(I - W)^T$ is a $n \times n$ matrix.

To make this a well-posed optimization problem, we solve it subject to the constraints that:

1. $\sum_{i=0}^{n} y_i = 0$

2. $\frac{1}{n} YY^T = I$, where I is a $p \times p$ identity matrix

As the solution to this constrained optimization problem, we compute the bottom $p + 1$ eigenvectors of M, i.e., the eigenvectors corresponding to the $p + 1$ lowest eigenvalues. Here the lowest eigenvalue is zero and the eigenvector corresponding to that is a unit vector. Hence, we discard the bottommost (constant) eigenvector and pick the remaining p eigenvectors from the bottom $p + 1$ vectors as the solution to this optimization problem, which is significantly different from Isomap which derives the output from the top eigenvectors [4]. This is how LLE finds the p embedding coordinates.

We know that LLE identifies the local geometry and produces neighborhood preserving mapping. However, when the number of nearest neighbors is greater than the input dimension ($k > d$), the least squares problem for finding the local weights is ill-conditioned and hence does not have a unique solution. Thus, a regularization term chosen relative to the trace of the local weight matrix is added to the reconstruction cost. However, if the choice of the regularization parameter is not appropriate, the regularized solution does not guarantee the optimal solution, which results in distorted embeddings. To address this regularization problem in LLE, Modified LLE that uses multiple local weight vectors was introduced [5]. In this method, multiple weight vectors were used for each data point in the reconstruction of the low dimensional embedding that retrieves the underlying geometry of the manifold without distortion.

9.2   ADVANTAGES AND LIMITATIONS

Using LLE for nonlinear dimensionality reduction is fairly simple and easy and has various advantages.

- LLE efficiently identifies the structure of the underlying manifold and very well preserves the local distances between data points of the original high dimensional space.
- LLE is highly efficient in discovering the nonlinear structures in the data and preserving the distances within the local neighborhood by computing neighborhood preserving embeddings.
- Another advantage of LLE is that it has only one free parameter, k, which is the number of nearest neighbors.
- Moreover, optimizations in LLE do not involve local minima.
- Lastly, it is computationally efficient as the computations involved in LLE are relatively faster since this approach tends to accumulate sparse matrices, thus greatly reducing the computational space and time associated with it.

While LLE has many advantages, it has some limitations as well.

- One major weakness of LLE is the regularization problem that results in distorted embeddings when the number of neighbors is greater than the input dimension. However, this problem is addressed by the Modified LLE algorithm that uses multiple weight vectors for reconstruction of each point.
- Apart from this, the fact that LLE is highly sensitive to noise and outliers in the data is a major limitation of this method.
- Another major drawback of this algorithm is that it assumes that the data lie on or near a smooth manifold, though it might not be true in some scenarios such as in multi-class classification problems. In such cases where the data has varying density and does not lie on a smooth manifold, LLE gives poor results.
- Also, as the k-nearest neighbors are chosen based on the Euclidean distances between the points, it is possible that two data points that actually do not lie on the same locally linear patch on the manifold might be grouped as

neighbors if the neighborhood distance is greater than the distance between the local patches on the manifold, leading to the problem of short-circuiting.
- Moreover, while LLE accurately preserves the local structure, it is less accurate in preserving the global structure as compared to Isomap.

9.3 USE CASES

Being an efficient nonlinear dimensionality reduction technique for high dimensional nonlinear data, LLE is widely used in face image processing and text data processing. The authors of the original paper [1] that proposed this algorithm illustrated the application of LLE on images of lips used in audio-visual speech synthesis. LLE is also applied to human motion data. Despite the configuration space being high dimensional due to various aspects such as variation in viewpoints or illumination, human motion and facial expressions intrinsically lie on low dimensional manifolds. LLE is used to learn this intrinsically low dimensional manifold in high dimensional motion data from wearable motion sensors for application in human activity recognition [6]. But its sensitivity to noise limits its application in many other areas. Moreover, variants of the LLE algorithm have been proposed by extending the original algorithm for various applications including gene expression data analysis [7], MRI-based disease classification, tomography reconstruction, cancer detection from Magnetic Resonance Spectroscopy (MRS), and anomaly detection [8].

9.4 EXAMPLE AND TUTORIAL

Example 1

To demonstrate how LLE works as a manifold learning technique, let us consider an illustration of dimensionality reduction on an S-curve dataset. This toy dataset has 2000 data points lying on an S-shaped manifold. The objective of manifold learning is to learn the underlying intrinsic geometry of the manifold and unfold the manifold in low dimensional space by preserving the locality of the data points.

For this example, we use sklearn to create an S-curve dataset with 2000 data points. Likewise, we use the *LocallyLinearEmbedding* algorithm implementation from the *sklearn,manifold* module. Finally, to plot the data, we use matplotlib.

Firstly, import all the required libraries.

```
from sklearn.manifold import LocallyLinearEmbedding
from sklearn import datasets
import matplotlib.pyplot as plt
from matplotlib.ticker import NullFormatter
```

Next, load the S-curve dataset with 2000 data points *(n_points)* using *datasets. make_s_curve.*

```
n_points = 2000
X, color = datasets.make_s_curve(n_points, random_state=0)
```

Now, create a 3D plot of the S-curve dataset.

```
fig = plt.figure(figsize=(45, 25))
ax = fig.add_subplot(251, projection='3d')
ax.scatter(X[:, 0], X[:, 1], X[:, 2], c=color, cmap=plt.
cm.jet, s=9, lw=1)
ax.view_init(10, -72)
```

Figure 9.1 is the visualization of the S-curve dataset on a three-dimensional scatter plot.

Next, we use LLE with number of neighbors as 10 ($k = 10$) and dimensionality of the target projection space *(n_components)* as 2. Here we use the standard LLE algorithm *(method = 'standard')*

```
lle= LocallyLinearEmbedding(n_neighbors=10, n_components=2,met
hod='standard')
X_lle= lle.fit_transform(X)
```

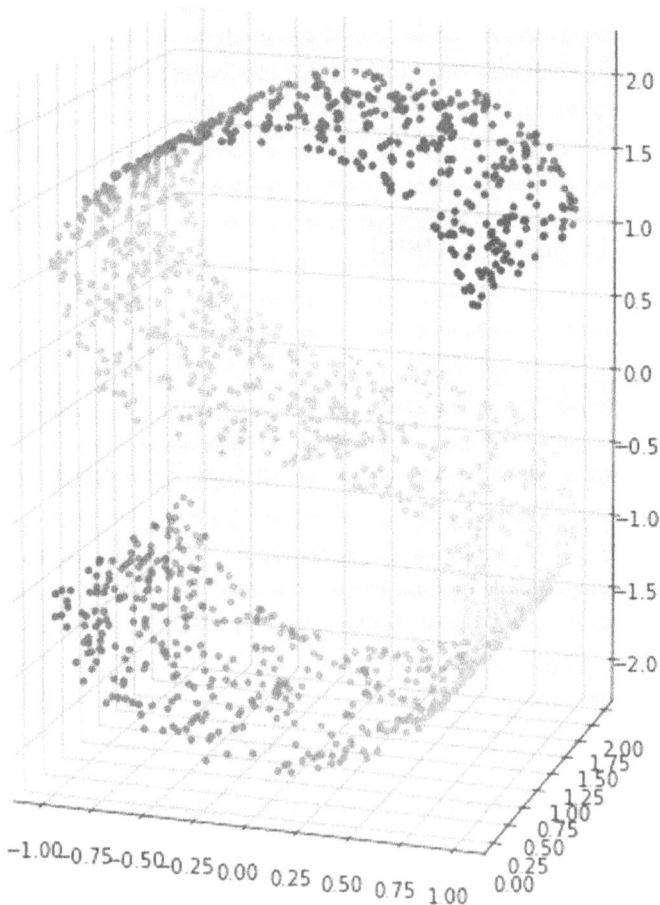

FIGURE 9.1   S-shaped curve, n=2000.

After fitting and transforming the data, plot the results.

```
fig = plt.figure(figsize=(5, 5))
ax = fig.add_subplot(1,1,1)
ax.scatter(X_lle[:, 0], X_lle[:, 1], c=color, cmap=plt.cm.jet,
s=9, lw=1)
ax.xaxis.set_major_formatter(NullFormatter())
ax.yaxis.set_major_formatter(NullFormatter())
ax.axis('tight')
plt.ylabel('Y coordinate')
plt.xlabel('X coordinate')
plt.show()
```

Figure 9.2 illustrates the result of manifold learning by LLE on the S-curve dataset with 2000 points and 10 neighbors ($k = 10$).

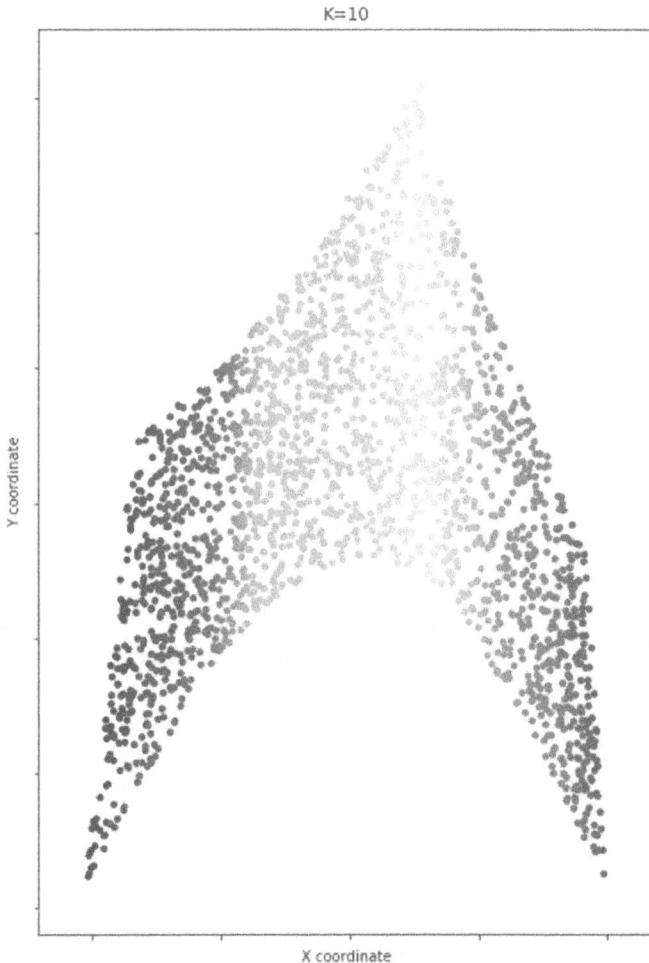

FIGURE 9.2   LLE (k=10).

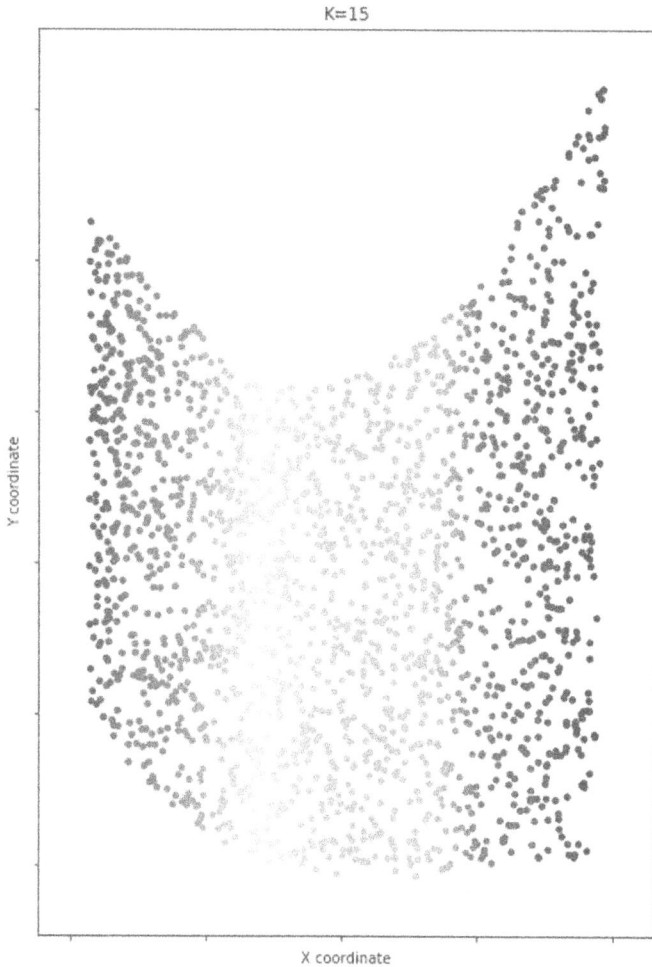

FIGURE 9.3   LLE (k=15).

Figure 9.3 is the result of dimensionality reduction by LLE with $k = 15$ (change to $n\_neighbors = 15$). Note how the resultant mappings differ with the choice of the number of neighbors k.

On the other hand, Figure 9.4 is the visualization of the result of manifold learning by the modified LLE algorithm on the same S-curve dataset with 10 nearest neighbors by changing parameter *method* to *'modified'* (*method = 'modified'*).

Example 2

Now that we have seen how LLE performs manifold learning on a toy dataset, let us visualize the results of dimensionality reduction by LLE on an MNIST hand-written digits dataset embedded in a 784-dimensional space. This dataset has

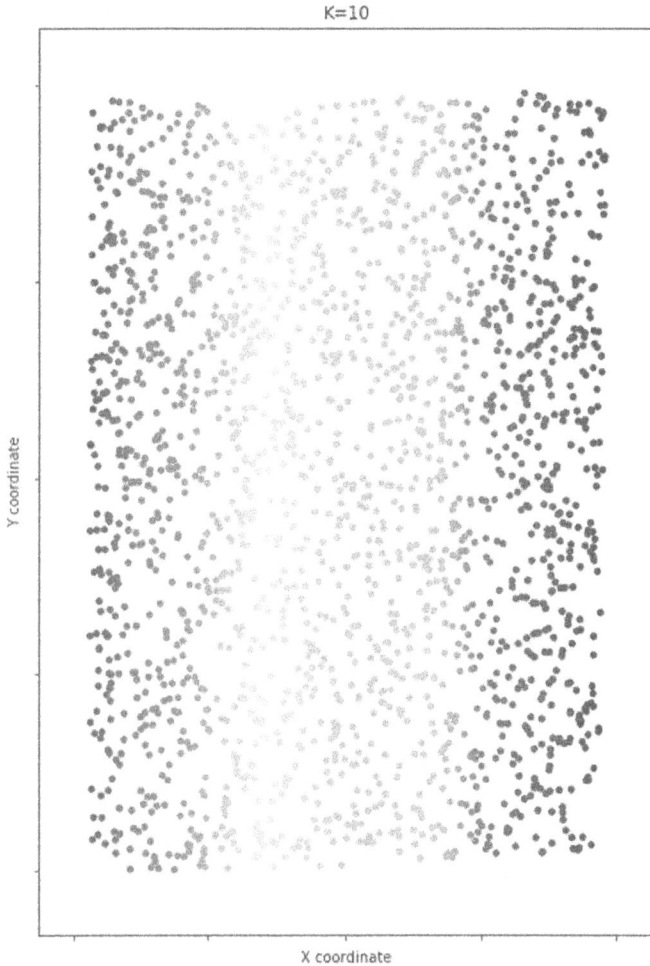

FIGURE 9.4 Modified LLE (k=10).

55000 data points where each data point is an image of size 28 × 28 pixels. Thus, the dimensionality of each data sample is 784. In this example, we will use LLE to reduce it to two dimensional.

We will use the MNIST handwritten digits dataset by importing it from the tensorflow library. Similar to the previous example, we will use the *sklearn.manifold* module from the scikit-learn library for implementing the LLE algorithm and finally plot the data using the matplotlib library.

Import all the necessary libraries.

```
import tensorflow as tf
from tensorflow.examples.tutorials.mnist import input_data
import sklearn
from sklearn.manifold import LocallyLinearEmbedding
import matplotlib.pyplot as plt
```

Load the dataset. X_train contains the images of digits and y_train contains the corresponding labels for each image.

```
mnist = input_data.read_data_sets("MNIST_data/")
X_train = mnist.train.images
y_train = mnist.train.labels
```

Use the LLE algorithm on the MNIST data to reduce its dimensionality. Here, we set the hyperparameters as follows:

- *n_neighbors* = *10* (Number of neighbors)
- *n_components* = 2 (dimensionality of the target projection space)

```
lle = LocallyLinearEmbedding(n_neighbors=10, n_components=2,
method='standard')
lle.fit(X_train)
X_lle = GaussianRP.transform(X_train)
```

Finally, plot the transformed data using matplotlib.

```
plt.figure(figsize=(10,10))
plt.scatter(X_lle[y_train==0, 0], X_lle[y_train==0, 1],
color='blue', alpha=0.5,label='0', s=9, lw=2)
plt.scatter(X_lle[y_train==1, 0], X_lle[y_train==1, 1],
color='purple', alpha=0.5,label='1',s=9, lw=2)
plt.scatter(X_lle[y_train==2, 0], X_lle[y_train==2, 1],
color='yellow', alpha=0.5,label='2',s=9, lw=2)
plt.scatter(X_lle[y_train==3, 0], X_lle[y_train==3, 1],
color='black', alpha=0.5,label='3',s=9, lw=2)
plt.scatter(X_lle[y_train==4, 0], X_lle[y_train==4, 1],
color='gray', alpha=0.5,label='4',s=9, lw=2)
plt.scatter(X_lle[y_train==5, 0], X_lle[y_train==5, 1],
color='lightblue', alpha=0.5,label='5', s=9, lw=2)
plt.scatter(X_lle[y_train==6, 0], X_lle[y_train==6, 1],
color='red', alpha=0.5,label='6',s=9, lw=2)
plt.scatter(X_lle[y_train==7, 0], X_lle[y_train==7, 1],
color='green', alpha=0.5,label='7',s=9, lw=2)
plt.scatter(X_lle[y_train==8, 0], X_lle[y_train==8, 1],
color='turquoise', alpha=0.5,label='8',s=9, lw=2)
plt.scatter(X_lle[y_train==9, 0], X_lle[y_train==9, 1],
color='orange', alpha=0.5,label='9',s=9, lw=2)
plt.ylabel('Y coordinate')
plt.xlabel('X coordinate')
plt.legend()
plt.show()
```

Figure 9.5 shows the visualization of the two-dimensional embedding of the MNIST dataset after dimensionality reduction by LLE with 10 neighbors (k=10).

On the other hand, Figure 9.6 shows the low dimensional representation of the MNIST data produced by the Modified LLE algorithm (change to *method* =

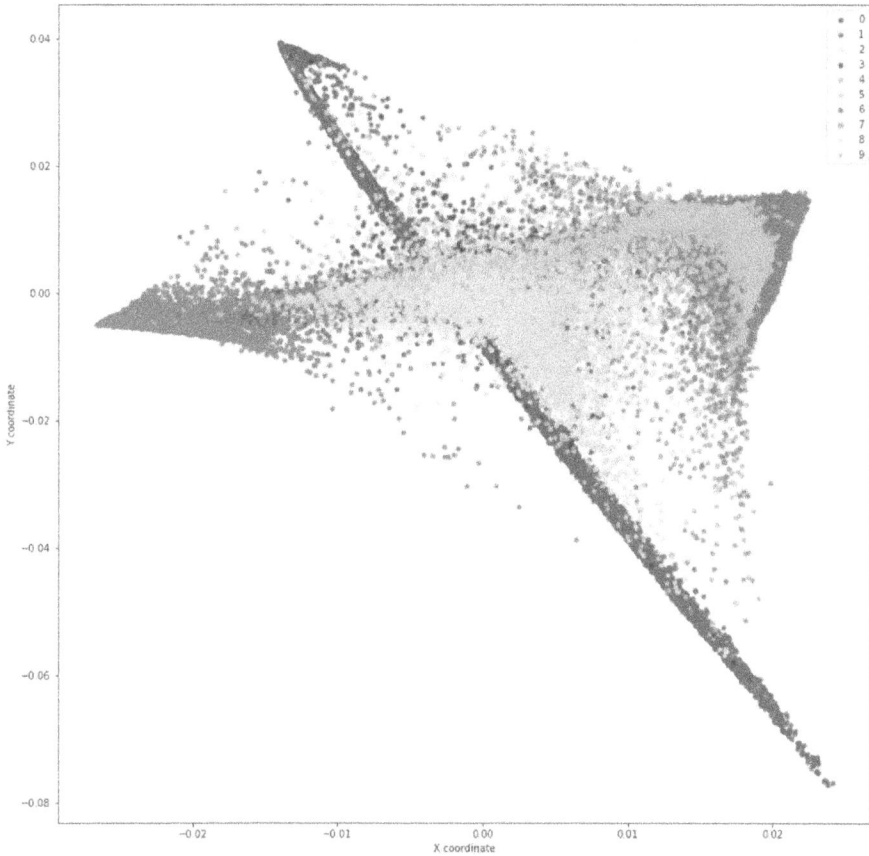

FIGURE 9.5   LLE on MNIST data (k=10).

'modified') with 10 neighbors (k=10). We can see the large overlaps between the digit classes in the mappings.

Example 3

Next, let us see an example of how LLE performs on the 64-dimensional digits data consisting of 1083 data points (images of size 8 × 8 pixels) that are embedded in a 64-dimensional space (8 × 8) and visualize the results. The data contains six classes of images each representing a digit from 0 to 5.

Let us import the digits dataset from the sklearn library. Similar to the previous example, we will use the *sklearn.manifold* module from the scikit-learn library for dimensionality reduction and matplotlib for plotting the data.

First, import the required libraries.

```
import sklearn
from sklearn import datasets
```

```
from sklearn.random_projection import GaussianRandomProjection
from sklearn.random_projection import SparseRandomProjection
import matplotlib.pyplot as plt
```

Load the data using *load_digits*.

```
digits = datasets.load_digits(n_class=6)
X = digits.data
y = digits.target
n_samples, n_features = X.shape
print(n_features)
print(n_samples)
```

Output:

```
64
1083
```

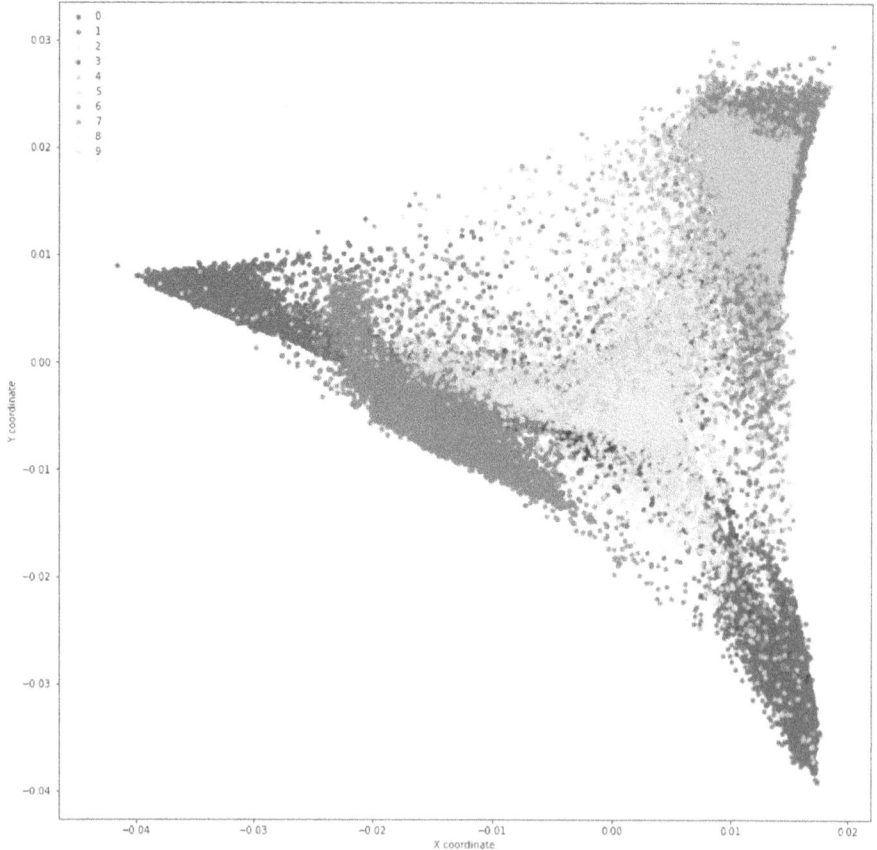

FIGURE 9.6 Modified LLE on MNIST data (k=10).

Use *sklearn.manifold.LocallyLinearEmbedding* to reduce the dimensionality of the data to two *(n_components = 2)*. Set parameter *n_neighbors = 30* (number of neighbors) and *method = 'Standard'* (for the standard LLE algorithm).

```
lle = LocallyLinearEmbedding(n_neighbors=30, n_components=2,
method='standard')
lle.fit(X)
X_lle = GaussianRP.transform(X)
```

Now, plot the reduced data.

```
plt.figure(figsize=(10,10))
plt.scatter(X_lle[y==0, 0], X_lle[y==0, 1], color='blue',
alpha=0.5, label='0', s=9, lw=1)
plt.scatter(X_lle[y==1, 0], X_lle[y==1,1], color='green',
alpha=0.5, label='1', s=9, lw=1)
```

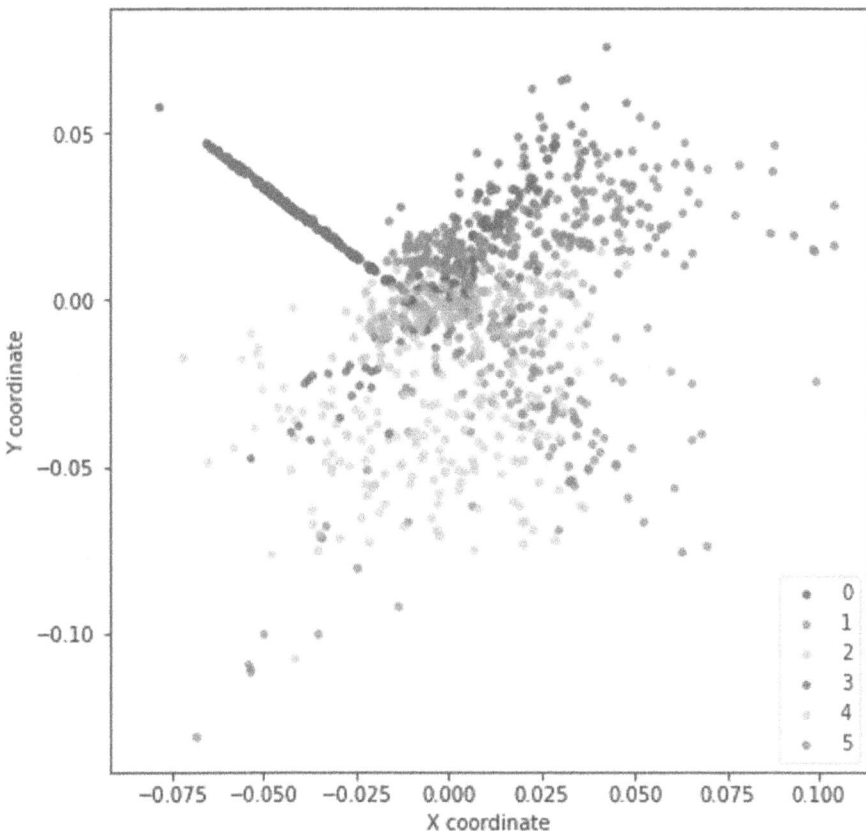

FIGURE 9.7   LLE on digits data (k=30).

```
plt.scatter(X_lle[y==2, 0], X_lle[y==2,1], color='orange',
alpha=0.5, label='2', s=9, lw=1)
plt.scatter(X_lle[y==3, 0], X_lle[y==3,1], color='purple',
alpha=0.5, label='3', s=9, lw=1)
plt.scatter(X_lle[y==4, 0], X_lle[y==4,1], color='violet',
alpha=0.5,label='4', s=9, lw=1)
plt.scatter(X_lle[y==5, 0], X_lle[y==5, 1], color='red',
alpha=0.5, label='5',s=9, lw=1)
plt.ylabel('Y coordinate')
plt.xlabel('X coordinate')
plt.legend()
plt.show()
```

Figure 9.7 represents the two-dimensional embedding of the data obtained from dimensionality reduction by the standard LLE algorithm with k=30.

Figure 9.8 is the visualization of the result by modified LLE algorithm (set parameter *method = 'modified'*) with k=30.

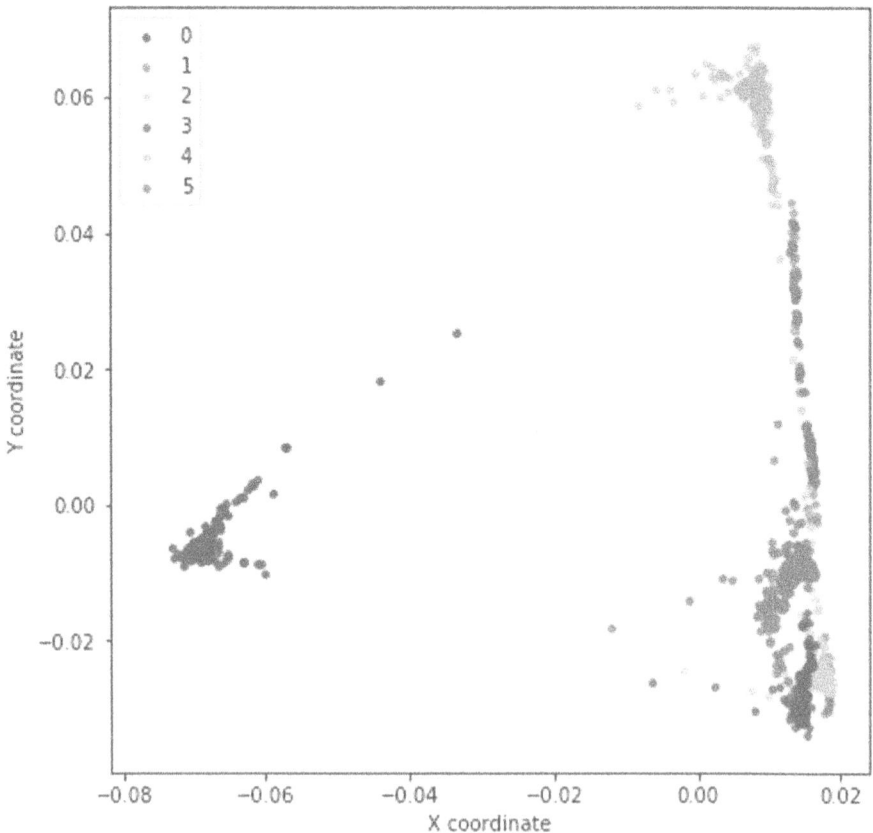

FIGURE 9.8   Modified LLE on digits data (k=30).

REFERENCES

1. Roweis, S. T., & Saul, L. K. (2000). Nonlinear dimensionality reduction by locally linear embedding. *Science*, *290*(5500), 2323–2326.
2. Saul, L. K., & Roweis, S. T. (2003). Think globally, fit locally: Unsupervised learning of low dimensional manifolds. *Journal of machine learning research*, *4*(Jun), 119–155.
3. Shalizi, C. (2009). Nonlinear dimensionality reduction I: Local linear embedding. *Data mining class*. Carnigie Mellon University (CMU), Pittsburgh.
4. Saul, L. K., Weinberger, K. Q., Sha, F., Ham, J., & Lee, D. D. (2006). Spectral methods for dimensionality reduction. *Semi-supervised learning*, *3*, pp. 293–308.
5. Zhang, Z., & Wang, J. (2007). MLLE: Modified locally linear embedding using multiple weights. In *Advances in neural information processing systems* (pp. 1593–1600). Vancouver, British Columbia, Canada.
6. Zhang, M., & Sawchuk, A. A. (2011, December). Manifold learning and recognition of human activity using body-area sensors. In *2011 10th International conference on machine learning and applications and workshops*, *2*, 7–13. IEEE, Washington, DC.
7. Pillati, M., & Viroli, C. (2005, June). Supervised locally linear embedding for classification: an application to gene expression data analysis. In *Proceedings of 29th annual conference of the German Classification Society (GfKl 2005)* (pp. 15–18). Magdeburg, Germany.
8. Ma, L., Crawford, M. M., & Tian, J. (2010). Anomaly detection for hyperspectral images based on robust locally linear embedding. *Journal of infrared, millimeter, and terahertz waves*, *31*(6), 753–762.

10 Spectral Clustering

10.1 EXPLANATION AND WORKING

The need to identify groups exhibiting similarity within their data for exploratory data analysis in many areas of application is the motivation behind various clustering techniques. While traditional clustering algorithms such as k-means or single linkage are widely used, spectral clustering has its own advantages. In this section we will discuss a clustering technique called spectral clustering, which by itself is not a dimensionality reduction technique but is closely related to another dimensionality reduction method called Laplacian Eigenmap [1] which will be discussed in the next chapter.

The intuitive goal of clustering is to cluster data into groups in a way that data points in the same group exhibit high similarity between them whereas data points belonging to different groups are dissimilar. To achieve this, given n points, x_i where i ranges from 1 to n, and a notion of similarity, we construct an undirected weighted graph $G = (V, E)$ called the similarity graph, where the vertices v_i represent the data points x_i and the edges represent the similarity between the data points. The weight of an edge tells us about the similarity between the two data points connected by that edge. If an edge has a positive weight (or larger than a certain threshold) associated with it, then the two data points that it connects are similar. On the other hand, if the weight of an edge is negative (or lower than the threshold), the two data points connected by that edge are dissimilar. The objective of clustering is to partition this graph in such a way that the weights of the edges between data points within the same group are maximized whereas the edges between data points belonging to different groups have very minimal weight. We will discuss in detail how spectral clustering can be derived in relation to this graph partitioning problem.

For two disjoint sets of data points A and $\bar{A}$, where $V = A \cup \bar{A}$, we define:

$$cut\left(A, \bar{A}\right) = \sum_{i \in A, \, j \in \bar{A}} w_{ij} \tag{10.1}$$

where w_{ij} represents the weight of the edge between a data point belonging to A and another data point from $\bar{A}$. For a given similarity graph, we can cluster the data points into k groups by solving the following min-cut problem:

$$cut\left(A_1, \ldots, A_k\right) = \sum_{i=1}^{k} cut\left(A_i, \bar{A}_i\right) \tag{10.2}$$

where $\bar{A}$ is the complement of A.

But the problem here is that this optimization problem yields a solution such that an individual vertex is separated from the rest of the graph, leading to bad clustering where a cluster has just one data point. To overcome this problem, we use the RatioCut [2], which is defined as follows:

$$RatioCut\left(A, \bar{A}\right) = \frac{cut\left(A, \bar{A}\right)}{|A|} + \frac{cut\left(\bar{A}, A\right)}{|\bar{A}|} \qquad (10.3)$$

$$RatioCut\left(A_1, ..., A_k\right) = \sum_{i=1}^{k} \frac{cut\left(A_i, \bar{A}_i\right)}{|A_i|} \qquad (10.4)$$

In RatioCut, the size of the cluster A is measured by the number of vertices it has, $|A|$. Thus, minimizing the RatioCut partitions the graph into balanced clusters in terms of the number of vertices in each cluster. Hence, our goal is to solve the following optimization problem.

$$\min_{A} RatioCut\left(A, \bar{A}\right)$$

We define a vector $f = (f_1, ..., f_n)^T \in \mathbb{R}^n$ such that:

$$f_i = \begin{cases} \sqrt{\dfrac{|\bar{A}|}{|A|}}, & \text{if } v_i \in A \\[3mm] -\sqrt{\dfrac{|A|}{|\bar{A}|}}, & \text{if } v_i \in \bar{A} \end{cases} \qquad (10.5)$$

Each vertex v_i is assigned a value f_i based on the group it belongs to. Now, minimizing RatioCut is equivalent to minimizing:

$$\sum_{ij} w_{ij} \left(f_i - f_j\right)^2$$

Proof:

$$\sum_{ij} w_{ij} \left(f_i - f_j\right)^2 = \sum_{i \in A, \, j \in \bar{A}} w_{ij} \left(\sqrt{\frac{|\bar{A}|}{|A|}} + \sqrt{\frac{|A|}{|\bar{A}|}}\right)^2 + \sum_{i \in \bar{A}, \, j \in A} w_{ij} \left(-\sqrt{\frac{|\bar{A}|}{|A|}} - \sqrt{\frac{|A|}{|\bar{A}|}}\right)^2$$

$$\text{[By (10.5)]}$$

$$= \sum_{i \in A, \, j \in \bar{A}} w_{ij} \left(\frac{|\bar{A}|}{|A|} + \frac{|A|}{|\bar{A}|} + 2\right) + \sum_{i \in \bar{A}, \, j \in A} w_{ij} \left(\frac{|\bar{A}|}{|A|} + \frac{|A|}{|\bar{A}|} + 2\right)$$

$$= \left(\frac{|\bar{A}|}{|A|} + \frac{|A|}{|\bar{A}|} + 2\right) \left(\sum_{i \in A, \, j \in \bar{A}} w_{ij} + \sum_{i \in \bar{A}, \, j \in A} w_{ij}\right)$$

$$= \left(\frac{|\bar{A}| + |A|}{|A|} + \frac{|A| + |\bar{A}|}{|\bar{A}|} \right) 2 \, cut\left(A, \bar{A}\right)$$

$$= 2|V| \left(\frac{1}{|A|} + \frac{1}{|\bar{A}|} \right) cut\left(A, \bar{A}\right)$$

[Since, $|A| + |\bar{A}| = |V|$. Here, $|V|$ is nothing but the total number of vertices, that is, the number of data points, n]:

$$2|V| \left(\frac{cut\left(A, \bar{A}\right)}{|A|} + \frac{cut\left(A, \bar{A}\right)}{|\bar{A}|} \right)$$

$$= 2|V| \, RatioCut\left(A, \bar{A}\right)$$

Therefore, minimizing $RatioCut\left(A, \bar{A}\right)$ is equivalent to minimizing $\sum_{ij} w_{ij} \left(f_i - f_j\right)^2$.

Now, $\sum_{ij} w_{ij} \left(f_i - f_j\right)^2$ can be expressed in terms of the Graph Laplacian as $f^T L f$.

Graph Laplacians are widely used for various clustering and graph partitioning problems [3–5]. Especially, the concept of Graph Laplacians is an important notion behind spectral clustering algorithms [6].

$$\sum_{ij} w_{ij} \left(f_i - f_j\right)^2 = f^T L f \tag{10.6}$$

Here, L is the Graph Laplacian matrix that is symmetric positive semidefinite defined as:

$$L = D - W \tag{10.7}$$

where W is an $n \times n$ weight matrix that captures the pairwise similarity between all the points and D is an $n \times n$ diagonal matrix in which each diagonal entry d_i is the summation of entries of the ith row of W, that is, $d_i = \sum_j w_{ij}$.

Proof:

$$f^T L f = f^T \left(D - W\right) f$$

$$= f^T D f - f^T W f$$

$$= \sum_i d_i f_i^2 - \sum_{ij} W_{ij} f_i f_j$$

$$= \frac{1}{2} \left(2 \sum_i d_i f_i^2 - 2 \sum_{ij} W_{ij} f_i f_j \right)$$

$$= \frac{1}{2}\left(\sum_i d_i f_i^2 + \sum_i d_i f_i^2 - 2\sum_{ij} W_{ij} f_i f_j \right)$$

$$= \frac{1}{2}\left(\sum_i f_i^2 \sum_j w_{ij} + \sum_i f_i^2 \sum_j w_{ij} - 2\sum_{ij} W_{ij} f_i f_j \right)$$

$$= \frac{1}{2}\left(\sum_{ij} w_{ij} f_i^2 + \sum_{ij} w_{ij} f_i^2 - 2\sum_{ij} W_{ij} f_i f_j \right)$$

$$= \frac{1}{2}\sum_{ij} w_{ij} \left(f_i - f_j \right)^2$$

Thus, to partition the graph, we solve the following simplified objective function which is equivalent to minimizing the RatioCut.

$$\min_f f^T L f \text{ subject to the constraint } f^T f = 1 \tag{10.8}$$

where matrix L is known and f is the solution to the problem which is nothing but the eigenvector corresponding to the second smallest eigenvalue of L. Here the smallest eigenvalue of L is zero and the eigenvector corresponding to it is a constant unit vector. Hence, we ignore it and take the next eigenvector to be the solution to this relaxed optimization problem. This real-valued n-dimensional vector f obtained as the solution determines the graph partitions as follows.

$$v_i \in \begin{cases} A, & \text{if } f_i \geq 0 \\ \bar{A}, & \text{if } f_i < 0 \end{cases}$$

That is, if f_i is positive then vertex v_i belongs to cluster A; otherwise it belongs to cluster $\bar{A}$. In this way, we yield the spectral clustering algorithm as an approximation to graph partitioning. Similarly, spectral clustering can be extended to more than two clusters (k>2), by taking the $n \times k$ matrix where the columns are the k eigenvectors corresponding to the smallest k non-zero eigenvalues and the ith row represents the ith node, v_i. Then, the k-means clustering method is used to cluster these n points into k clusters.

10.2   ADVANTAGES AND LIMITATIONS

- Spectral clustering is a powerful method that is simple to implement and outperforms traditional clustering techniques in many cases. One of the main advantages of spectral clustering is that it does not make any strong assumptions on the shape of the clusters, unlike K-means clustering which assumes that the points belonging to a cluster are spherical about the cluster center while that might not be true. Hence, in such cases, spectral clustering results in more accurate clusters of arbitrary shapes.

- On the other hand, while this approach is efficient as long as the similarity graph is sparse, it becomes computationally expensive when the data is large and the graph is dense. As compared to other clustering techniques such as k-means or center-based clustering methods which need to compute and store only the distances between the points from its centers, spectral clustering is relatively computationally expensive since it has to deal with similarities between all pairs of data points and also perform eigendecomposition which is a computation-intensive step unless the matrix is sparse.
- Another drawback of the spectral clustering method is that it is affected by the presence of noise and outliers in the data. An outlier which has very low similarity with the other data points produces an eigenvalue very close to 1, making the principal eigenvector to be an outlier and not cluster. Hence, outliers have to be removed before performing eigendecomposition.
- Another issue with the spectral clustering technique is the choice of similarity graph. Choosing a good similarity measure is important as spectral clustering is not consistent with different choices of the similarity graphs and their parameters (parameter k in the case of k-nearest neighbor graph and parameter ε-neighborhood graph) [6]. Therefore, careful consideration has to be given in choosing the appropriate parameters in order to produce good results.

10.3   USE CASES

Clustering techniques are widely used for exploratory data analysis with a wide range of applications. In almost every field that deals with empirical data such as statistics, data mining, machine learning, and pattern recognition, analysis of data by identifying groups within data that exhibit similar behavior is very common, and this is where clustering techniques come into the picture. Spectral clustering methods often out-perform traditional clustering algorithms in many aspects, and hence they are widely preferred for applications in genomics for tasks such as clustering of protein sequences, VLSI design, in computer vision and image processing applications especially for image segmentation [5], in speech processing, in text classification and other natural language processing applications, and various computer vision problems. The advantages of spectral clustering methods over other traditional clustering techniques such as the k-means clustering algorithm make it suitable for a wide range of applications.

10.4   EXAMPLES AND TUTORIAL

Example 1

To demonstrate how spectral clustering works, let us apply spectral clustering on the MNIST handwritten digits database to cluster the classes of digits. Since the data is high dimensional, for the purpose of visualizing the clustering results, let us first reduce the dimensionality of the data by mapping it to a two-dimensional space using any dimensionality reduction technique and then perform spectral clustering on the reduced data.

In this example, let us use Isomap for reducing the dimensions of the data for clustering and visualization. For this let us use *sklearn.manifold.Isomap* from the *sklearn.manifold* module of the scikit-learn library for implementing the Isomap algorithm. For the dataset, we will import the MNIST handwritten digits dataset from the tensorflow library. To plot the reduced data, we will use matplotlib library. After reducing the dimensionality of the data using Isomap, we will perform spectral clustering on the reduced data. For this, we will use *sklearn.cluster.SpectralClustering* from the *sklearn.cluster* module of the scikit-learn library. Finally, let us plot the clustered data using matplotlib.

Firstly, import all the necessary libraries.

```
import tensorflow as tf
from tensorflow.examples.tutorials.mnist import input_data
from sklearn.manifold import Isomap
from sklearn.cluster import SpectralClustering
import matplotlib.pyplot as plt
import matplotlib.colors
```

Next, let us load the MNIST dataset. Here X contains the digits images and y contains the corresponding labels.

```
mnist = input_data.read_data_sets("MNIST_data/")
X = mnist.train.images
y = mnist.train.labels
X_train=X[:5000]
y_train= y[:5000]
X_test=X[10000:15000]
y_test= y[10000:15000]
```

Let us use the Isomap algorithm on the imported dataset to reduce the dimensions. Let the number of neighbors *(n_neighbors)* be 5 and the dimensionality of the target space *(n_components)* be 2.

```
iso = Isomap(n_neighbors=5, n_components=2)
iso.fit(X_train)
X_iso = iso.transform(X_train)
```

Now, let us visualize the reduced data using matplotlib.

```
plt.figure(figsize=(10,10))
plt.scatter(X_iso[y_train==0, 0], X_iso[y_train==0, 1],
color='blue', alpha=0.5,label='0', s=9, lw=2)
plt.scatter(X_iso[y_train==1, 0], X_iso[y_train==1, 1],
color='purple', alpha=0.5,label='1',s=9, lw=2)
plt.scatter(X_iso[y_train==2, 0], X_iso[y_train==2, 1],
color='yellow', alpha=0.5,label='2',s=9, lw=2)
plt.scatter(X_iso[y_train==3, 0], X_iso[y_train==3, 1],
color='black', alpha=0.5,label='3',s=9, lw=2)
plt.scatter(X_iso[y_train==4, 0], X_iso[y_train==4, 1],
color='gray', alpha=0.5,label='4',s=9, lw=2)
```

```
plt.scatter(X_iso[y_train==5, 0], X_iso[y_train==5, 1],
color='turquoise', alpha=0.5,label='5',s=9, lw=2)
plt.scatter(X_iso[y_train==6, 0], X_iso[y_train==6, 1],
color='red', alpha=0.5,label='6',s=9, lw=2)
plt.scatter(X_iso[y_train==7, 0], X_iso[y_train==7, 1],
color='green', alpha=0.5,label='7',s=9, lw=2)
plt.scatter(X_iso[y_train==8, 0], X_iso[y_train==8, 1],
color='violet', alpha=0.5,label='8',s=9, lw=2)
plt.scatter(X_iso[y_train==9, 0], X_iso[y_train==9, 1],
color='orange', alpha=0.5,label='9',s=9, lw=2)
plt.ylabel('Y coordinate')
plt.xlabel('X coordinate')
plt.legend()
plt.show()
```

Figure 10.1 is the visualization of the result of dimensionality reduction on the 784-dimensional MNSIT data (having 5000 data points) using Isomap.

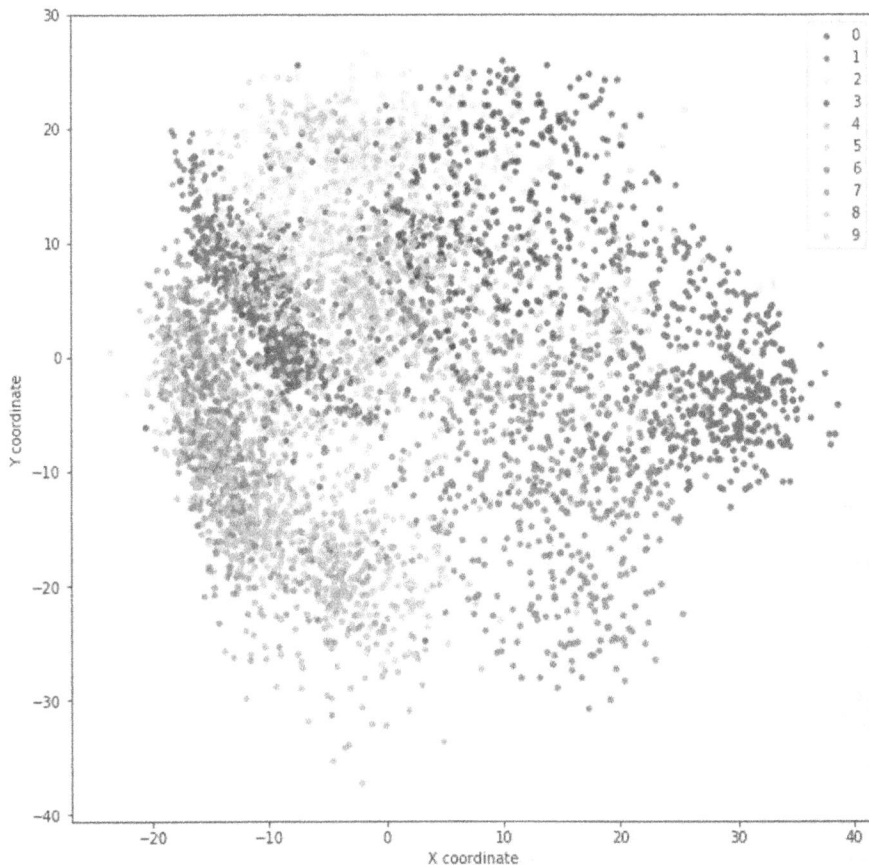

FIGURE 10.1   Two-dimensional representation of MNIST data.

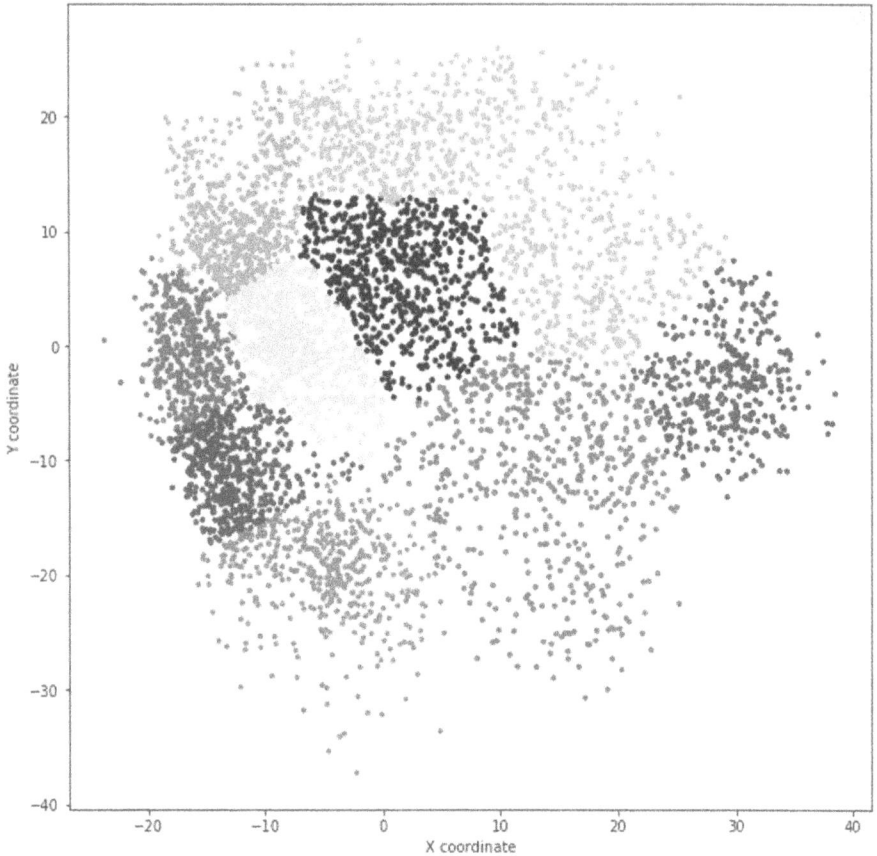

FIGURE 10.2   Result of spectral clustering on two-dimensional MNIST data.

Following this, spectral clustering is performed on this data to group the data points into ten clusters (digits 0–9).

```
spectral = SpectralClustering(n_clusters=10, eigen_
solver='arpack', affinity="nearest_neighbors")
X = spectral.fit(X_iso)
y_pred = spectral.fit_predict(X_iso)
```

Let us visualize the clustered data using matplotlib.

```
cmap = matplotlib.colors.LinearSegmentedColormap.from_list("",
['blue','purple','yellow','black','gray','turquoise','red',
'green','violet','orange'])
plt.figure(figsize=(10,10))
#color= ['blue','purple','yellow','black','gray','turquoise','
red','green','violet','orange']
i=['0','1','2','3','4','5','6','7','8','9']
```

```
plt.scatter(X_iso[:, 0], X_iso[:, 1], c=y_pred, s=9, lw=1,
cmap=cmap)
plt.ylabel('Y coordinate')
plt.xlabel('X coordinate')
plt.legend()
plt.show()
```

The resulting clusters are visualized as shown in Figure 10.2. It can be noted that spectral clustering has resulted in different classes being grouped into clusters with a clear separation between them.

REFERENCES

1. Belkin, M., & Niyogi, P. (2002). Laplacian eigenmaps and spectral techniques for embedding and clustering. In *Advances in neural information processing systems* (pp. 585–591). Vancouver, British Columbia, Canada.
2. Hagen, L., & Kahng, A. B. (1992). New spectral methods for ratio cut partitioning and clustering. *IEEE transactions on computer-aided design of integrated circuits and systems, 11*(9), 1074–1085.
3. Belkin, M., & Niyogi, P. (2003). Laplacian eigenmaps for dimensionality reduction and data representation. *Neural computation, 15*(6), 1373–1396.
4. Ng, A. Y., Jordan, M. I., & Weiss, Y. (2002). On spectral clustering: Analysis and an algorithm. In *Advances in neural information processing systems* (pp. 849–856). Vancouver, British Columbia, Canada.
5. Shi, J., & Malik, J. (2000). Normalized cuts and image segmentation. *IEEE transactions on pattern analysis and machine intelligence, 22*(8), 888–905.
6. Von Luxburg, U. (2007). A tutorial on spectral clustering. *Statistics and computing, 17*(4), 395–416.

11 Laplacian Eigenmap

11.1 EXPLANATION AND WORKING

In this section we will discuss a computationally efficient approach to non-linear dimensionality reduction called the Laplacian Eigenmap, which has a natural connection to the spectral clustering method that we discussed in Chapter 10 [1]. As the name suggests, this algorithm uses the notion of Graph Laplacians to compute a low dimensional, neighborhood preserving representation of the original data [2]. While Graph Laplacians are extensively used for graph partition and clustering problems, this algorithm uses Graph Laplacian matrices for the problem of dimensionality reduction [2–4]. Unlike dimensionality reduction techniques like PCA and MDS, Laplacian Eigenmaps take into consideration that the structure of the underlying manifold the low dimensional representation generated by this algorithm closely reflects the intrinsic geometric structure of the manifold by preserving the neighborhood information from high dimensional space by constructing weighted neighborhood graphs. This algorithm is structurally similar to Locally Linear Embedding (LLE) which has a connection to the Laplacian [2, 5, 6].

Given n points $x_1, ..., x_n \in \mathbb{R}^d$, we construct a weighted graph $G = (V, E)$ with n vertices (data points) and a set of edges connecting each vertex with its neighbors. An edge is created between two vertices v_i and v_j if they are neighbors. This neighborhood relation can be established in one of the following ways [2]:

1. ε-*neighborhood* – Two vertices x_i and x_j are connected if the pairwise Euclidean distance between them is lesser than ε, where $\varepsilon \in \mathbb{R}$ is a parameter.
2. *K-nearest neighborhood* – Vertex x_i is a neighbor of x_j and is connected to it by an edge if and only if x_i is among the k nearest neighbors of x_j.

Next, we assign weights to edges of the graph. Weights can be defined in the following two ways.

1. Define $w_{ij} = \begin{cases} e^{\frac{-|x_i - x_j|^2}{t}}, & \text{if } x_i \text{ and } x_j \text{ are connected by an edge} \\ 0, & \text{otherwise} \end{cases}$

   where t is a real valued parameter called the heat kernel.

2. Define $w_{ij} = \begin{cases} 1, & \text{if } x_i \text{ and } x_j \text{ are connected by an edge} \\ 0, & \text{otherwise} \end{cases}$

Let $Y = \begin{bmatrix} y_1 & y_2 & \cdots & y_n \end{bmatrix}^T \in \mathbb{R}^p$ be an $n \times p$ matrix where the ith row gives the mapping of the ith vertex on a p-dimensional space. We define a cost function:

$$\sum_{ij} w_{ij} |y_i - y_j|^2 = Tr(Y^T LY) \tag{11.1}$$

where y_i is the p-dimensional representation of the ith vertex and L is the Graph Laplacian matrix. This cost function gives a high penalty if any two points x_i and x_j that are neighbors in the original d-dimensional space are mapped far apart on the p-dimensional space, that is, if y_i and y_j are chosen to be far apart. When w_{ij} is large, we minimize the objective function by minimizing $|y_i - y_j|^2$, i.e., by choosing y_i and y_j to be close. Thus, minimizing this objective function preserves the locality as it ensures that if two points in the high dimensional space appear close to each other, they should be close to each other on the low dimensional space as well:

$$\min_{y} \sum_{ij} w_{ij} |y_i - y_j|^2$$

$$= \min_{Y} Tr(Y^T LY) \text{ subject to the constraint } Y^T DY = I$$

Introducing this constraint to avoid collapse onto a subspace of dimension less than p–1. Using the Lagrangian, the solution to this constrained optimization problem is given by the matrix of eigenvectors corresponding to the p smallest non-zero eigenvalues of the generalized eigenvalue problem $Ly = \lambda Dy$.

As we can see, this optimization problem is very similar to that of the spectral clustering method. Interestingly, we observe that the locality preserving dimensionality reduction method yields the same solution as spectral clustering. By this we realize that this approach to dimensionality reduction, in which mappings by the eigenvectors of the Graph Laplacian are used, is closely related to spectral clustering methods discussed in Chapter 10.

Another noteworthy observation is the Locally Linear Embedding (LLE) algorithm's connection to the Laplacian. As we discussed in Chapter 9, LLE attempts to minimize $Y(I - W)(I - W)^T Y^T$ by solving for the eigenvectors of $(I - W)(I - W)^T$ which can be interpreted as finding the eigenfunctions of the Laplacian.

11.2   ADVANTAGES AND LIMITATIONS

- Laplacian Eigenmap is an efficient algorithm that preserves the intrinsic geometric structure of the underlying manifold in a low dimensional space. It is also computationally efficient for sparse neighborhood graphs as local methods involve only a sparse eigenvalue problem. Thus, Laplacian Eigenmaps can be scaled to large datasets.
- Also, there is no local optima and hence only one unique solution.
- Laplacian Eigenmap is relatively insensitive to noise and outliers because of its property to preserve locality [1].

- Since only the local distances are used, it is not susceptible to the problem of short circuiting.
- However, it is challenging to choose the right parameters for the Laplacian Eigenmap as it is extremely parameter sensitive. Since there is no defined way to choose the heat kernel parameter t, it is critical to choose the appropriate t in order to obtain good representations in low dimensional space [2]. Similarly, the effect of different choices of the parameters k (in the k-neighborhood graph) and ε (in the ε-neighborhood graph) on the behavior of the embeddings is not fully understood. Hence, selecting the appropriate parameter for this algorithm is not trivial.

11.3   USE CASES

Laplacian Eigenmaps are used for nonlinear dimensionality reduction in a wide range of application areas for their locality preserving properties. Since it is a computationally efficient method for representing high dimensional data and has a natural connection to clustering, it is preferred for non-linear dimensionality reduction of complex data involved in application areas like information retrieval and data mining. Laplacian Eigenmaps are applied to naturally occurring complex datasets with very high dimensionality in the domains of vision, speech, and language to obtain low dimensional representations of the data for many practical applications. For example, they are used to construct low dimensional representation of text data for search and information retrieval from documents and also to group words that belong to similar syntactic categories. Another example is the application of Laplacian Eigenmaps to human speech where the speech signals are of high dimensions, whereas the distinct phonetic dimensions in speech are less. Here, Laplacian Eigenmaps are used to create low dimensional representations of high dimensional speech signals such that they are correlated with the phonetic content.

11.4   EXAMPLES AND TUTORIAL

Example 1

Let us illustrate this algorithm on some datasets. First, let us take some toy datasets such as S-curve and Swiss roll and visualize how Laplacian Eigenmaps perform manifold learning on these datasets.

In this example, let us take the S-curve dataset. This toy dataset has 2000 data points lying on an S-shaped manifold. We use sklearn to create an S-curve dataset with 2000 data points. Likewise, we use *sklearn.manifold.SpectralEmbedding* from the scikit-learn library which provides the implementation of Laplacian Eigenmap. Finally, to plot the data, we use matplotlib.

Firstly, import all the required libraries.

```
from sklearn.manifold import SpectralEmbedding
from sklearn import datasets
import matplotlib.pyplot as plt
from matplotlib.ticker import NullFormatter
```

Next, load the S-curve dataset with 2000 data points *(n_points)* using *datasets.make_s_curve.*

```
n_points = 2000
X, color = datasets.make_s_curve(n_points, random_state=0)

fig = plt.figure(figsize=(45, 25))
ax = fig.add_subplot(251, projection='3d')
ax.scatter(X[:, 0], X[:, 1], X[:, 2], c=color, cmap=plt.
cm.jet, s=9, lw=1)
ax.view_init(10, -72)
```

Let us create a 3D plot of the S-curve dataset. Figure 11.1 is the scatter plot of the S-curve data with 2000 data points.

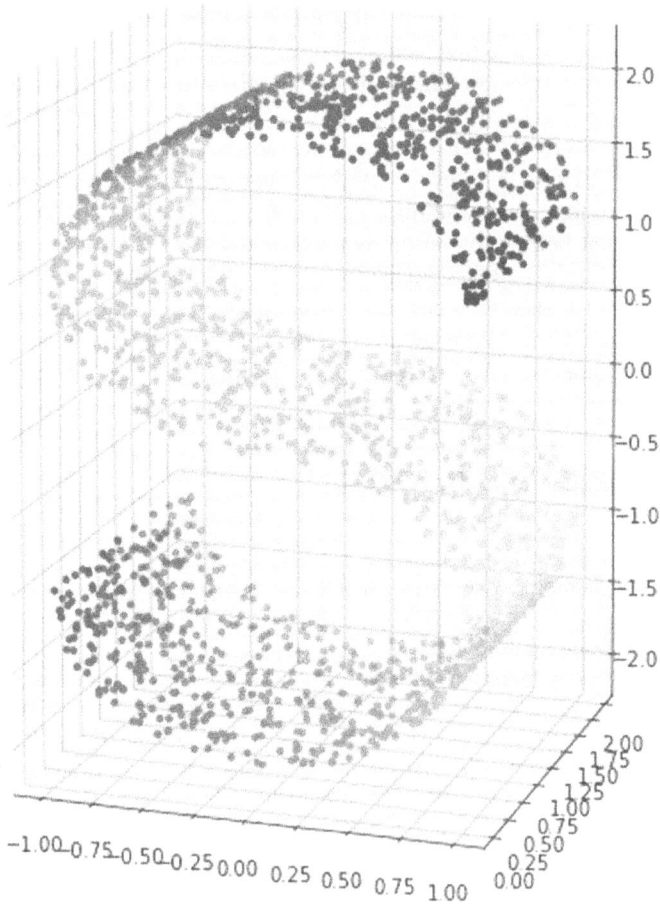

FIGURE 11.1   S-curve (n=2000).

Next, we use the Laplacian Eigenmap with dimensionality of the target projection space *(n_components)* as 2 and number of neighbors *(n_neighbors)* for constructing the graph as 10.

```
se= SpectralEmbedding (n_components=2, n_neighbor = 10)
X_se= se.fit_transform(X)
```

After fitting and transforming the data, let us plot the results.

```
fig = plt.figure(figsize=(5, 5))
ax = fig.add_subplot(1,1,1)
ax.scatter(X_se[:, 0], X_se[:, 1], c=color, cmap=plt.cm.jet,
s=9, lw=1)
ax.xaxis.set_major_formatter(NullFormatter())
ax.yaxis.set_major_formatter(NullFormatter())
ax.axis('tight')
plt.ylabel('Y coordinate')
plt.xlabel('X coordinate')
plt.show()
```

Figure 11.2 is the visualization of the S-curve data on a two-dimensional space after dimensionality reduction using the Laplacian Eigenmap.

Example 2

Now, let us take consider an example using the Swiss roll dataset. The Swiss roll in this example contains 2000 data points. Similar to the previous example, let us apply Laplacian Eigenmaps using *sklearn,manifold.SpectralEmbedding* from the scikit-learn library.
First let us import all the required libraries.

```
from sklearn.manifold import SpectralEmbedding
from sklearn import datasets
import matplotlib.pyplot as plt
from matplotlib.ticker import NullFormatter
```

Next, load the S-curve dataset with 2000 data points *(n_points)* using *datasets. make_swiss_roll.*

```
n_points = 2000
X, color = datasets.make_swiss_roll(n_points, random_state=0)
```

Create a 3D plot of the Swiss roll dataset using matplotlib.

```
fig = plt.figure(figsize=(45, 25))
ax = fig.add_subplot(251, projection='3d')
ax.scatter(X[:, 0], X[:, 1], X[:, 2], c=color, cmap=plt.
cm.jet, s=9, lw=1)
ax.view_init(10, -72)
```

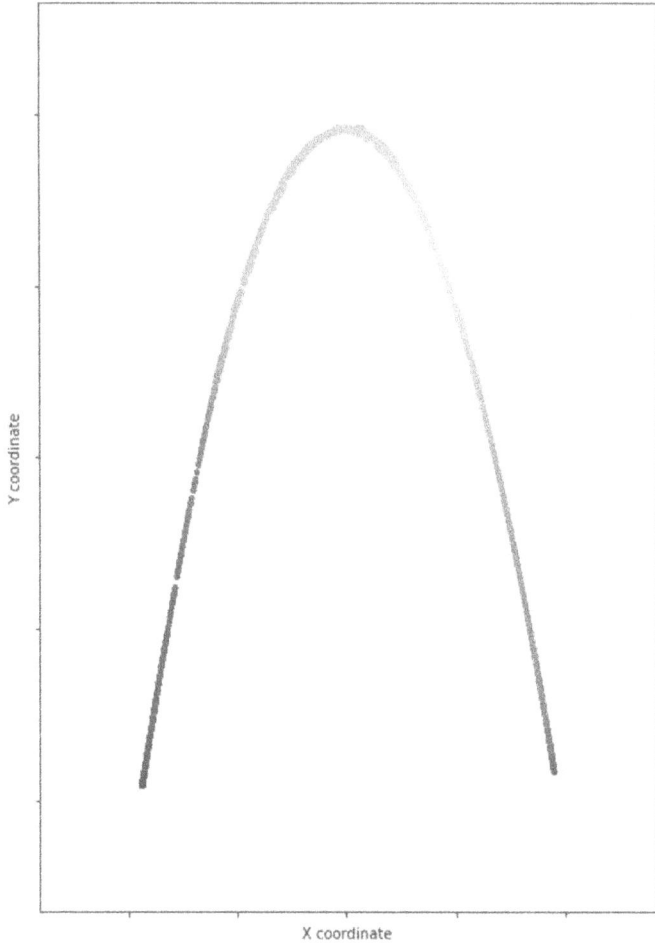

FIGURE 11.2   Result of Laplacian Eigenmap on the S-curve data.

Figure 11.3 is the scatter plot of Swiss roll data consisting of 2000 data points.

Next, we use the Laplacian Eigenmap to reduce this dataset to two dimensions with the number of neighbors *(n_neighbors)* for constructing the graph as 10.

```
se= SpectralEmbedding (n_components=2, n_neighbor = 10)
X_se= se.fit_transform(X)
```

After fitting and transforming the data, let us plot the results.

```
fig = plt.figure(figsize=(5, 5))
ax = fig.add_subplot(1,1,1)
ax.scatter(X_se[:, 0], X_se[:, 1], c=color, cmap=plt.cm.jet,
s=9, lw=1)
ax.xaxis.set_major_formatter(NullFormatter())
ax.yaxis.set_major_formatter(NullFormatter())
```

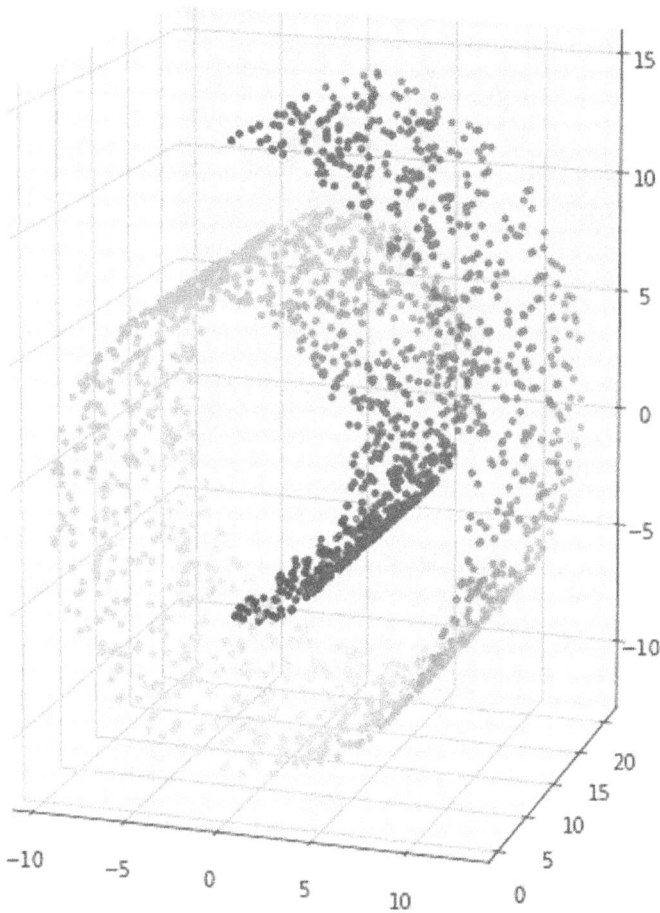

FIGURE 11.3   Swiss roll (n=2000).

```
ax.axis('tight')
plt.ylabel('Y coordinate')
plt.xlabel('X coordinate')
plt.show()
```

Figure 11.4 is the representation of the data in the low dimensional space.

It can be noted that unlike Isomap, Laplacian Eigenmap does not isometrically map the data points onto a two-dimensional space. But it attempts to unfold the manifold, that is, unroll the S-curve and the Swiss roll, thereby preserving the locality of the data points on the manifold, though the distances are not preserved.

Example 3

Now that we have seen how the Laplacian Eigenmap performs dimensionality reduction on a toy dataset, let us illustrate the results of this dimensionality reduction technique on some real-world datasets. Let us take the MNIST handwritten

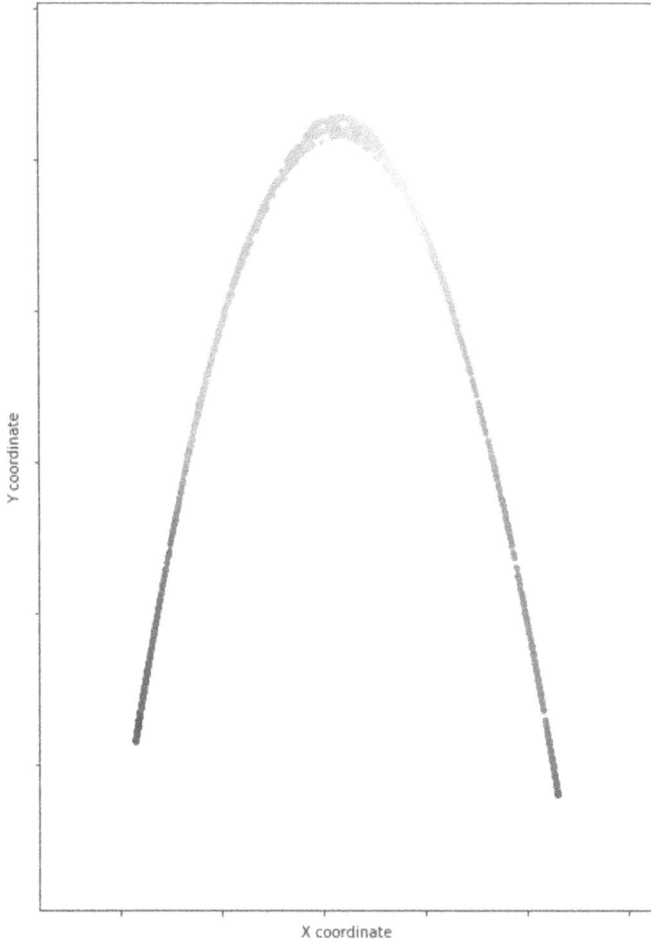

FIGURE 11.4   Result of Laplacian Eigenmap on the Swiss roll data.

database consisting of 60,000 data samples of dimensionality 784, out of which we will use 2000 sample points on which we perform dimensionality reduction using the Laplacian Eigenmap.

We will use the MNIST handwritten digits dataset by importing it from the tensorflow library. Similar to the previous example, we will use *sklearn.manifold. SpectralEmbedding* from scikit-learn's *sklearn.manifold* module for implementing the Laplacian Eigenmap algorithm and finally plot the data using matplotlib library.

Import all the required libraries.

```
import tensorflow as tf
from tensorflow.examples.tutorials.mnist import input_data
import sklearn
from sklearn.manifold import SpectralEmbedding
import matplotlib.pyplot as plt
```

Load the MNIST dataset. X_train contains the images of digits and y_train contains the corresponding labels for each image.

```
mnist = input_data.read_data_sets("MNIST_data/")
X_train = mnist.train.images
y_train = mnist.train.labels
```

Use the Laplacian Eigenmap to reduce this dataset to two dimensions with 10 as the number of neighbors *(n_neighbors)* for constructing the graph.

```
se = SpectralEmbedding(n_neighbors=10, n_components=2,method=
'standard')
se.fit(X_train)
X_se = se.transform(X_train)
```

Finally, plot the resultant data using matplotlib.

```
plt.figure(figsize=(10,10))
plt.scatter(X_se[y_train==0, 0], X_se[y_train==0, 1],
color='blue', alpha=0.5,label='0', s=9, lw=2)
plt.scatter(X_se[y_train==1, 0], X_se[y_train==1, 1],
color='purple', alpha=0.5,label='1',s=9, lw=2)
plt.scatter(X_se[y_train==2, 0], X_se[y_train==2, 1],
color='yellow', alpha=0.5,label='2',s=9, lw=2)
plt.scatter(X_se[y_train==3, 0], X_se[y_train==3, 1],
color='black', alpha=0.5,label='3',s=9, lw=2)
plt.scatter(X_se[y_train==4, 0], X_se[y_train==4, 1],
color='gray', alpha=0.5,label='4',s=9, lw=2)
plt.scatter(X_se[y_train==5, 0], X_se[y_train==5, 1],
color='lightblue', alpha=0.5,label='5',s=9, lw=2)
plt.scatter(X_se[y_train==6, 0], X_se[y_train==6, 1],
color='red', alpha=0.5,label='6',s=9, lw=2)
plt.scatter(X_se[y_train==7, 0], X_se[y_train==7, 1],
color='green', alpha=0.5,label='7',s=9, lw=2)
plt.scatter(X_se[y_train==8, 0], X_se[y_train==8, 1],
color='turquoise', alpha=0.5,label='8',s=9, lw=2)
plt.scatter(X_se[y_train==9, 0], X_se[y_train==9, 1],
color='orange', alpha=0.5,label='9',s=9, lw=2)
plt.ylabel('Y coordinate')
plt.xlabel('X coordinate')
plt.legend()
plt.show()
```

Figure 11.5 illustrates the two-dimensional representation of the data after dimensionality reduction by Laplacian Eigenmap (using k-neighborhood graph).

Example 4

In this example, we use the 64-dimensional digits dataset with 1063 data samples in six classes (digits 0–5). Let us import this dataset from the scikit-learn library.

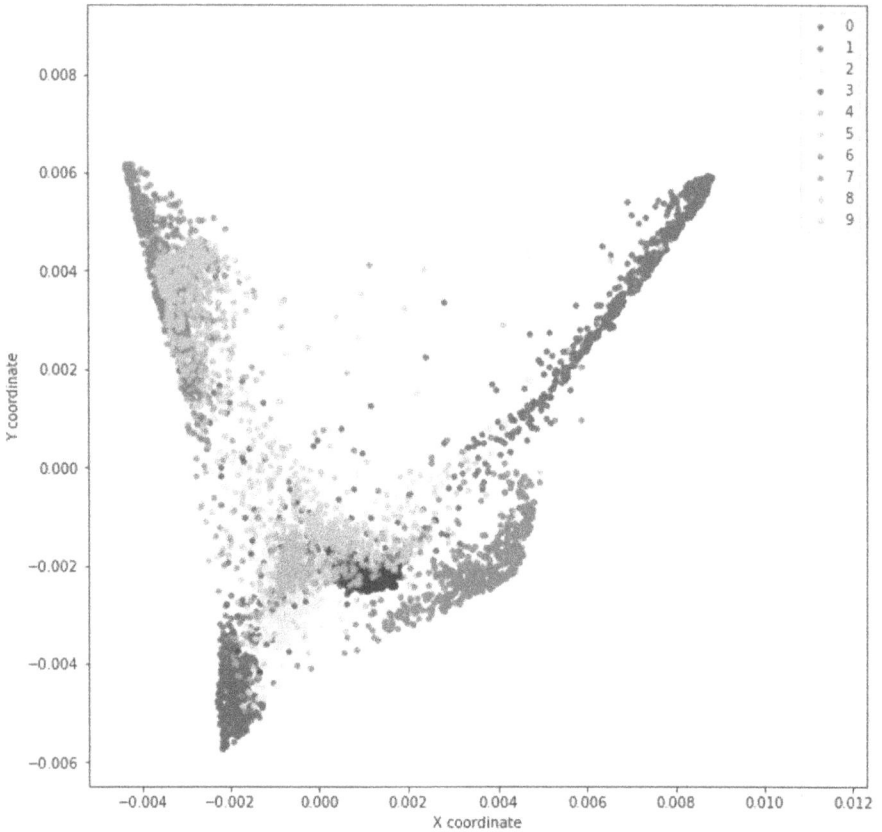

FIGURE 11.5   Visualization of result of Laplacian Eigenmap on MNIST data.

To do that, first let us import all the necessary libraries.

```
import sklearn
from sklearn import datasets
from sklearn.manifold import SpectralEmbedding
import matplotlib.pyplot as plt
```

Load the dataset using *load_digits*.

```
digits = datasets.load_digits(n_class=6)
X = digits.data
y = digits.target
```

Use *sklearn.manifold.SpectralEmbedding* to reduce the dimensionality of the data to two *(n_components = 2)*.

```
se = SpectralEmbedding(n_neighbors=10, n_components=2,method=
'standard')
se.fit(X)
X_se = se.transform(X)
```

Finally, plot the resultant data using matplotlib.

```
plt.figure(figsize=(10,10))
plt.scatter(X_se[y==0, 0], X_se[y==0, 1], color='blue',
alpha=0.5,label='0', s=9, lw=2)
plt.scatter(X_se[y==1, 0], X_se[y==1, 1], color='green',
alpha=0.5,label='1',s=9, lw=2)
plt.scatter(X_se[y==2, 0], X_se[y==2, 1], color='orange',
alpha=0.5,label='2',s=9, lw=2)
plt.scatter(X_se[y==3, 0], X_se[y==3, 1], color='purple',
alpha=0.5,label='3',s=9, lw=2)
plt.scatter(X_se[y==4, 0], X_se[y==4, 1], color='violet',
alpha=0.5,label='4',s=9, lw=2)
plt.scatter(X_se[y==5, 0], X_se[y==5, 1], color='red',
alpha=0.5,label='5',s=9, lw=2)
plt.ylabel('Y coordinate')
plt.xlabel('X coordinate')
plt.legend()
plt.show()
```

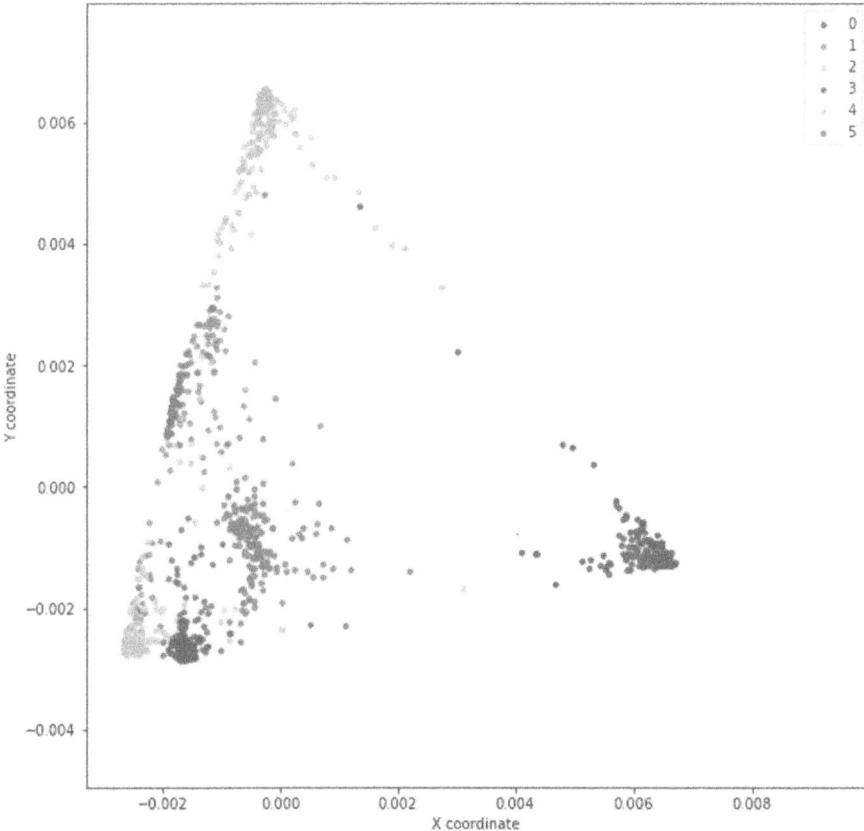

FIGURE 11.6   Visualization of result of Laplacian Eigenmap on digits data.

Figure 11.6 visualizes the result of Laplacian Eigenmap on the 64-dimensional hand-written digits dataset with 1063 data samples (digits 0–5). It is the low dimensional representation obtained by mapping the non-linear data from a high dimensional space to a two-dimensional space using the Laplacian Eigenmap algorithm.

REFERENCES

1. Belkin, M., & Niyogi, P. (2002). Laplacian eigenmaps and spectral techniques for embedding and clustering. In *Advances in neural information processing systems* (pp. 585–591). Vancouver, British Columbia, Canada.
2. Belkin, M., & Niyogi, P. (2003). Laplacian eigenmaps for dimensionality reduction and s representation. *Neural computation*, *15*(6), 1373–1396.
3. Ng, A. Y., Jordan, M. I., & Weiss, Y. (2002). On spectral clustering: Analysis and an algorithm. In *Advances in neural information processing systems* (pp. 849–856). Vancouver, British Columbia, Canada.
4. Shi, J., & Malik, J. (2000). Normalized cuts and image segmentation. *IEEE transactions on pattern analysis and machine intelligence*, *22*(8), 888–905.
5. Saul, L. K., Weinberger, K. Q., Sha, F., Ham, J., & Lee, D. D. (2006). Spectral methods for dimensionality reduction. *Semi-supervised learning*, *3*, pp. 293–308.
6. Roweis, S. T., & Saul, L. K. (2000). Nonlinear dimensionality reduction by locally linear embedding. *Science*, *290*(5500), 2323–2326.

12 Maximum Variance Unfolding

12.1 EXPLANATION AND WORKING

While the classical dimensionality reduction technique of Principal Component Analysis (PCA) is a powerful and a widely used method for linear dimensionality reduction, it fails in the case of non-linear data. However, in non-linear cases, PCA can still be adapted by exploiting the kernel trick. This is done by generalizing the classical PCA to non-linear cases by replacing a kernel function for the inner product in feature space [1]. This non-linear generalization of PCA is nothing but the kernel PCA that was discussed in Chapter 4. Though linear kernels, polynomial kernels, and Gaussian kernels are some of the widely used kernels, the choice of the kernel is very important as different kernels yield different low dimensional mappings. So how do we choose the kernel? While in some techniques like the kernel PCA we start from a fixed kernel, we can instead formulate an optimization problem from which the algorithm tries to learn the optimal kernel matrix that yields an embedding of the data points and that unfolds the underlying manifold in the feature space by semidefinite programming. This dimensionality reduction technique is Semidefinite Embedding (SDE) [2, 3] also called as Maximum Variance Unfolding (MVU).

In this section, we will discuss a semidefinite programming-based approach to non-linear dimensionality reduction where the algorithm learns a kernel matrix for reducing the dimensionality. This is a fundamentally different approach involving semidefinite programming [4] as compared to other manifold learning algorithms like LLE and Isomaps. In this approach, we set up a convex optimization problem to find an optimal kernel matrix that maps high dimensional data lying on a manifold to a lower dimensional feature space. The kernel mapping is obtained by maximizing the variance of the embeddings in the low dimensional space subject to constraints that preserve the distances and angles between the data points and their neighbors.

The objective of the algorithm is to learn a kernel matrix that unfolds the underlying manifold in a low dimensional space by capturing the similarity in a local patch, that is, by preserving the distance between each point with its nearest neighbors. This algorithm is based on a simple intuition that can be explained using an analogy. Imagine the data points x_i where $i = 1, \ldots, n$ to be steel balls that are connected to their nearest neighbors by rigid metallic rods. The algorithm attempts to maximize the variance by pulling apart the data points without breaking the rigid rods. In this way, the total variance is maximized while the distances and the angles between the data points do not change. This is exactly described by the term "Maximum

Variance Unfolding," where it tries to maximize the variance of its embeddings in low dimensional feature space, subject to the constraint that the distances and angles between data points and their neighbors are preserved. To put it in a nutshell, this algorithm takes advantage of semidefinite programming to learn a kernel matrix that gives embeddings in the feature space such that the variance of the embeddings is maximum and the distances are preserved. The optimization problem with the constraints and the objective function formulated to find the kernel are discussed in detail in the following section.

12.1.1 CONSTRAINTS ON THE OPTIMIZATION

The optimization over the kernel matrix is constrained by three conditions that have to be satisfied. The first criterion that has to be met is the positive semi-definiteness of the kernel matrix K. Thus, the kernel matrix should be a symmetric matrix with non-negative eigenvalues. That is, $K \geq 0$.

Next, we define a constraint on the kernel matrix K for a locally isometric mapping that preserves the distances and angles between the data points and their neighbors. Let $x_1, x_2, \ldots x_n \ \varepsilon \ \mathbb{R}^d$ be the original data points in d-dimensional space and $y_1, y_2, \ldots y_n \ \varepsilon \ \mathbb{R}^p$ be the embeddings in the feature space of dimension p (p<d). The kernel matrix K that is to be learned is defined by the inner product of the embeddings in the feature space as:

$$K_{ij} = \langle y_i, y_j \rangle \tag{12.1}$$

For all (i,j) such that x_i and x_j are k-nearest neighbors, the constraint to preserve the distance and angle between a data point and its neighbors is defined as:

$$\left| x_i - x_j \right|^2 = \left| y_i - y_j \right|^2 \tag{12.2}$$

$$\begin{aligned} \left| x_i - x_j \right|^2 &= y_i.y_i + y_j.y_j - 2y_i.y_j \\ &= K_{ii} + K_{jj} - 2K_{ij} \end{aligned} \qquad \text{[By (12.1)].}$$

This can be considered as the constraint imposed by the rigid rods to preserve the distances between the connected data points.

Finally, the embeddings in feature space are also constrained to be centered on the origin:

$$\sum_i y_i = 0 \tag{12.3}$$

which can be expressed in terms of the kernel matrix as

$$0 = \left| \sum_i y_i \right|^2 = \sum_{ij} y_i.y_i = \sum_{ij} K_{ij} \qquad \text{[By (12.1)].}$$

12.1.2 OBJECTIVE FUNCTION

As mentioned earlier, an objective function is optimized to learn a kernel matrix that gives a locally isometric mapping of the data points in the feature space. As already discussed in Section 12.1, this can be understood by imagining the underlying manifold to be analogous to the steel balls connected to its k-nearest neighbors by rigid rods that remain connected when pulled apart by locking the neighborhood and preserving the distance and angles between the neighbors and at the same time flattens the manifold, increasing the total variance. This intuition can be formalized as an optimization over positive semidefinite matrices by defining an objective function that measures the distance between the outputs, given by:

$$\frac{1}{2N} \sum_{ij} |y_i - y_j|^2 \tag{12.4}$$

This can be expressed in terms of the kernel matrix as

$$\frac{1}{2N} \sum_{ij} |y_i - y_j|^2 = \frac{1}{2N} \sum_{ij} K_{ii} + K_{jj} - 2K_{ij} = Tr(K)$$

Thus, the objective function for the optimization is the trace of the kernel matrix. By maximizing this objective function, that is, by maximizing the trace of the kernel matrix, the embeddings are pulled as far apart as possible, thus maximizing the variance in the feature space.

This objective function subject to the three constraints forms an optimization problem that is an instance of semidefinite programming as it is a convex optimization problem which attempts to optimize a linear function of the elements of a positive semidefinite matrix subject to constraints that are linear equalities. The optimization problem for MVU can be summarized as:

$$maxTr(K)$$

Subject to

1. $K \geq 0$

2. $\sum_{ij} K_{ij} = 0$

3. $|x_i - x_j|^2 = K_{ii} + K_{jj} - 2K_{ij}$ for all i, j where i and j are neighbors

Once the kernel matrix is computed by solving the optimization problem, the p-dimensional embeddings are obtained from the p leading eigenvectors corresponding to the largest p eigenvalues by singular value decomposition of the kernel matrix. Since the kernel matrix optimization in this approach involves semidefinite programming, this version of kernel PCA is referred to as Semidefinite Embedding (SDE) [2].

12.2   ADVANTAGES AND LIMITATIONS

- MVU or SDE is a nonlinear dimensionality reduction approach that over-comes certain limitations of other manifold learning techniques like LLE and Isomaps.
- MVU ensures that its constraints will result in locally isometric embed-dings. It very well maintains the distances and angles between the data points and their neighbors while precisely unfolding the underlying mani-fold in the feature space. The intrinsic dimensionality of the manifold is very well reflected by the learned kernel matrix.
- Another major advantage of SDE is that the optimization problem being convex eliminates the possibility of spurious local maxima, thus guarantee-ing a unique solution to the problem.
- Moreover, the flexibility to relax the distance-preserving constraints in the semidefinite programming problem make this algorithm more adaptable to handle the noise in the data.
- While SDE has its own advantages, a notable disadvantage of this algo-rithm is its computational complexity. As the size of the semidefinite program scales linearly with the size of the data, the time complexity of solving the semidefinite program for the n × n matrix is significantly high when the number of data points n is large. Hence, the major drawback of this approach relative to other manifold learning approaches is the time associated with solving problems in semidefinite programming in the case of large datasets making it difficult to efficiently scale SDE up to large datasets.
- However, a much faster algorithm for manifold learning by semidefinite programming called landmark SDE was developed to reduce the computa-tion time in the case of large datasets [5].
- Furthermore, while the performance of the optimized kernels learned by the SDE algorithm for problems in manifold learning is better than other kernels such as the polynomial or Gaussian kernel in kernel PCA, it is sig-nificantly worse in the case of large margin classification [6, 7] unless the decision boundary of the unfolded manifold is nearly linear [2].

12.3   USE CASES

The maximum variance unfolding algorithm can be adapted to specific applications by relaxing the distance-preserving constraints in the semidefinite programming to handle the noise in the data, for instance [8]. This algorithm is used in various computer vision problems and image processing applications including face image processing and recognition. In natural language processing, MVU is used to create a low dimensional representation of words while still preserving the semantic relation-ship between them. Moreover, MVU is used as a noise reduction method for non-linear signals where the noisy signals are embedded in a high dimensional space and then non-linear dimensionality reduction is performed using MVU to create a low dimensional representation; this low dimensional manifold is then reconstructed to obtain the noise-reduced signal. This MVU-based noise reduction method is applied in fault diagnosis. Similarly, MVU is used for the problem of dimensionality reduction in various practical applications.

REFERENCES

1. Schölkopf, B., Smola, A., & Müller, K. R. (1998). Nonlinear component analysis as a kernel eigenvalue problem. *Neural computation, 10*(5), 1299–1319.
2. Weinberger, K. Q., Sha, F., & Saul, L. K. (2004, July). Learning a kernel matrix for nonlinear dimensionality reduction. In *Proceedings of the twenty-first international conference on machine learning* (p. 106). Banff, Alberta, Canada.
3. Weinberger, K. Q., & Saul, L. K. (2006). Unsupervised learning of image manifolds by semidefinite programming. *International journal of computer vision, 70*(1), 77–90.
4. Vandenberghe, L., & Boyd, S. (1996). Semidefinite programming. *SIAM review, 38*(1), 49–95.
5. Weinberger, K. Q., Packer, B., & Saul, L. K. (2005, January). Nonlinear dimensionality reduction by semidefinite programming and kernel matrix factorization. In *AISTATS; Proceedings of the tenth international workshop on artificial intelligence and statistics*, Barbados.
6. Lanckriet, G. R., Cristianini, N., Bartlett, P., Ghaoui, L. E., & Jordan, M. I. (2004). Learning the kernel matrix with semidefinite programming. *Journal of machine learning research, 5*(Jan), 27–72.
7. Graepel, T. (2002). Kernel matrix completion by semidefinite programming. In *International conference on artificial neural networks* (pp. 694–699). Springer, Berlin, Heidelberg.
8. Weinberger, K. Q., & Saul, L. K. (2006, July). An introduction to nonlinear dimensionality reduction by maximum variance unfolding. In *AAAI; Proceedings of the twenty-first association for the advancement of artificial intelligence conference on artificial intelligence* (Vol. 6, pp. 1683–1686). Boston, MA.

13 t-Distributed Stochastic Neighbor Embedding (t-SNE)

13.1 EXPLANATION AND WORKING

We have discussed various non-linear dimensionality reduction techniques that preserve the local structure of the data in a low dimensional space. While these algorithms very well preserve the local geometry of the data, they do not retain both the local as well as the global structure of the data in a single mapping onto the low dimensional space. In this chapter, we discuss a relatively new dimensionality reduction technique called t-SNE which captures the local structure of the original data and at the same time, reveals the global structure at different scales [1]. t-SNE is a variant of Stochastic Neighbor Embedding (SNE) [2] which addresses some of the pitfalls of SNE. Firstly, let us discuss the SNE algorithm which forms the basis for t-SNE.

13.1.1 STOCHASTIC NEIGHBOR EMBEDDING (SNE)

SNE converts distances between data points to conditional probabilities. For any two data points x_i and x_j in high dimensional space d, the conditional probability $p_{j|i}$ is a measure of how likely it is for x_i to choose x_j as its neighbor if the neighbors were to be chosen based on their probability density under a Gaussian centered at x_i. This is as if, for any data point x_i, we put a Gaussian centered at x_i, and based on the density of this Gaussian we decide if other points are its neighbors or not. If two data points x_i and x_j are close in the original space, then $P_{j|i}$ will be relatively high, whereas if they are far apart, $p_{j|i}$ will be low. This conditional probability is given by

$$p_{j|i} = \frac{\exp\left(\frac{-\left|x_i - x_j\right|^2}{2\sigma_i^2}\right)}{\sum\limits_{k \neq i} \exp\left(\frac{-\left|x_i - x_k\right|^2}{2\sigma_i^2}\right)} \tag{13.1}$$

where σ_i is the variance of the Gaussian centered at x_i. This variance depends on the individual data point x_i, and it is not a unique variance defined over the whole space. Also, $P_{i|i}$ is set to zero as we are concerned only about pairwise similarities between the data points and hence avoid any point choosing itself as its neighbor.

Similarly, let y_i and y_j be the mapping of the original data points x_i and x_j respectively in the low dimensional space p. It is possible to compute a similar probability $q_{j|i}$ in the low dimensional space given by:

$$q_{j|i} = \frac{\exp\left(-\left|y_i - y_j\right|^2\right)}{\sum_{k \neq i} \exp\left(-\left|y_i - y_k\right|^2\right)} \tag{13.2}$$

Here the variance $\sigma = \frac{1}{\sqrt{2}}$ is constant throughout the low dimensional space. Also, $q_{i|i}$ is set to zero. If the similarity between the data points in the original space is retained in the low dimensional mapping as well, then the conditional probabilities in both the spaces, $p_{j|i}$ and $q_{j|i}$ will be equal. Hence, we need to choose a mapping $y_1, \ldots, y_n \in \mathbb{R}^p$ in the low dimensional space p, such that it minimizes the difference between $p_{j|i}$ and $q_{j|i}$. We define a cost function C, called a Kullback-Leibler divergence (KL divergence), as a measure of the distance between the two conditional probabilities. The objective of SNE is to find the low dimensional representation by minimizing this cost function, that is, by minimizing the sum of KL divergences over all the points. The cost function is defined as:

$$C = \sum_i KL(P_i \| Q_i) = \sum_{ij} p_{j|i} \log \frac{p_{j|i}}{q_{j|i}} \tag{13.3}$$

where, for a given point x_i and its mapping y_i, P_i and Q_i are the conditional probability distributions over all the other points and map points respectively. Note that $KL(P_i \| Q_i)$ is not the same as $KL(Q_i \| P_i)$, that is, KL divergence is not symmetric. Hence, the different types of mismatch in the distances between points when mapped onto a low dimensional space have different effects on the value of the cost function. If a high $p_{j|i}$ (x_i and x_j are neighbors) is modelled by a small $q_{j|i}$ (y_i and y_j are far apart), it incurs a high cost (since $p_{j|i} > q_{j|i}$ and log of a large value is large), whereas, if a low $p_{j|i}$ (x_i and x_j are far apart) is modelled by a high $q_{j|i}$ (y_i and y_j are neighbors), a low cost is incurred.

The optimization of the cost function becomes difficult, and moreover, SNE suffers from the "crowding problem." This crowding problem arises due to the fact that high dimensions have more space to accommodate data points than low dimensions. When we embed a data point from a high dimensional space to a lower dimensional space, the mapped points become overcrowded. Suppose, if $d + 1$ points can be accommodated equidistant from each other on a d dimensional space, then when we embed these points in a lower dimensional space p (where $p < d$), these points have to be squished or moved closer to each other as the area available to accommodate the points is reduced, leading to distortion of the distances between them. This is termed the crowding problem.

Thus, a new technique called t-distributed SNE (t-SNE) was introduced as a variant of SNE that aims at alleviating these drawbacks by making the cost function of SNE symmetric and using t-distribution instead of Gaussian distribution. This algorithm is different from SNE in two ways:

1. t-SNE uses symmetric joint probabilities where $p_{ij} = p_{ji}$ (in high dimension) and $q_{ij} = q_{ji}$ (in low dimension), unlike the asymmetric conditional probabilities, $p_{j|i}$ and $q_{j|i}$ used in SNE. Here, the pairwise similarity in high dimension in terms of probability, p_{ij} is given by:

$$p_{ij} = \frac{\exp\left(\dfrac{-\left|x_i - x_j\right|^2}{2\sigma^2}\right)}{\sum_{k \neq i} \exp\left(\dfrac{-\left|x_i - x_k\right|^2}{2\sigma^2}\right)} \tag{13.4}$$

2. In the high dimensional space, Gaussian distribution is used to convert distances into probabilities, whereas in the low dimensional mapping, rather than using Gaussian, t-SNE uses student t-distribution which has heavier tails than Gaussian. As heavy-tailed distribution allows more space for points in the low dimensional space, using t-distribution overcomes the crowding problem. The joint probability q_{ij} in low dimension using t-distribution is given by:

$$q_{ij} = \frac{\dfrac{1}{1 + \left|y_i - y_j\right|^2}}{\sum_{k \neq i}\left(\dfrac{1}{1 + \left|y_i - y_k\right|^2}\right)}$$

$$q_{ij} = \frac{\left(1 + \left|y_i - y_j\right|^2\right)^{-1}}{\sum_{k \neq i}\left(1 + \left|y_i - y_k\right|^2\right)^{-1}} \tag{13.5}$$

Thus, t-SNE finds the low dimensional mapping $y_1, \ldots, y_n \in \mathbb{R}^p$ by minimizing the symmetric cost function using the gradient descent method, that is, by minimizing the KL divergence between the symmetric joint probability distribution P in high dimension and the t-distribution based joint probability distribution Q in the low dimension, given by:

$$C = KL(P \| Q) = \sum_{ij} p_{ij} \log \frac{p_{ij}}{q_{ij}} \tag{13.6}$$

13.2   ADVANTAGES AND LIMITATIONS

- t-SNE outperforms other techniques in handling non-linear data efficiently as it very well captures the complex polynomial relationships between the features making it a highly efficient nonlinear dimensionality reduction technique.

- Also, as discussed earlier in Section 13.1, t-SNE very well preserves the locality of the data while at the same time reveals some of the important global structure of the data in the low dimensional mapping. It has the tendency to identify the structure of the data at many scales and also reveals high dimensional data that lie on multiple manifolds and clusters.
- Although t-SNE outperforms various other non-linear dimensionality reduction techniques by providing very good mappings in low dimensional space, it has a few potential drawbacks.
- While its performance is efficient on small datasets, its quadratic time and space complexity makes it computationally complex when applied on reasonably large datasets [1]. This makes it computationally very slow on datasets with more than a few thousand data points. To overcome this, the landmark approach to t-SNE was proposed that made it computationally efficient to visualize large real-world datasets.
- Additionally, the performance of t-SNE on general dimensionality reduction where the dimensionality of the data is reduced to a dimension greater than three (p>3) is not very clear. The performance of t-SNE when reducing to a two- or three-dimensional space cannot be generalized to dimensions more than three because of the heavy tails of the student t-distribution which, in high dimensional spaces, lead to mappings that do not preserve the local structure of the data as the heavy tails encompass a large proportion of the probability mass under the t-distribution in high dimensional spaces.
- Another major weakness of this algorithm is that it fails when the data has intrinsically high dimensional structure as t-SNE performs dimensionality reduction mainly based on the local properties of the data. t-SNE's assumption on the local linearity of the manifold is debased in the case of data having high intrinsic dimensionality and highly varying manifold, making it less successful on such data.
- Furthermore, another limitation is the non-convexity of its cost function. Since the cost function of t-SNE is not convex it is required to choose many optimization parameters and this affects the constructed solutions as they depend on the choice of the optimization parameters. Sometimes the same choice of parameters can give different results in different runs.
- However, experiments on various real-world datasets show that t-SNE outperforms various state-of-the-art dimensionality reduction and data visualization techniques in many ways.

13.3   USE CASES

t-SNE is a powerful data exploration and data visualization technique with various applications in image processing, bioinformatics, signal processing, speech processing, NLP, and many more. It is used to visualize high dimensional data by identifying patterns in data and mapping them into low dimensional space for effective visualization with diverse applications such as computational biology, computer security, music analysis, and cancer biology [3]. This method is extensively used in image processing applications like facial expression recognition and medical imaging and is also used for natural language processing problems like text comparison using word vector representations and finding word embeddings in low dimensional space. It is also applied on genomic data which are generally of very high dimensions for applications like gene sequence analysis [4].

13.4.   EXAMPLES

Example 1

Let us now visualize the results of dimensionality reduction by t-SNE on the 64-dimensional digits dataset and the MNIST digits dataset. Firstly, let us perform dimensionality reduction on the digits dataset with 1083 samples lying in a 64-dimensional space to map it to a two-dimensional feature space.

For this example, we use the *sklearn.manifold* module from the scikit-learn library which implements various manifold learning algorithms among which t-SNE is one.

First, import the necessary libraries.

```
import sklearn
from sklearn import datasets
from sklearn.manifold import TSNE
import matplotlib.pyplot as plt
```

Next, load the dataset from sklearn. There are 6 classes of digits and 1083 data points in this loaded dataset. Each data point is of size 8×8 pixels. Hence, the dimensionality is 64.

```
digits = datasets.load_digits(n_class=6)
X = digits.data
y = digits.target
n_samples, n_features = X.shape
print(n_features)
print(n_samples)
```

Output:

```
64
1083
```

Use the t-SNE model from sklearn where the parameter *n_components* denotes the dimensionality of the target projection space, which in this case is 2. Then fit and transform the data using t-SNE.

```
tsne = TSNE(n_components=2, random_state=0)
X_tsne = tsne.fit_transform(X)
```

Now, let us visualize the transformed data using matplotlib. The results are visualized in Figure 13.1.

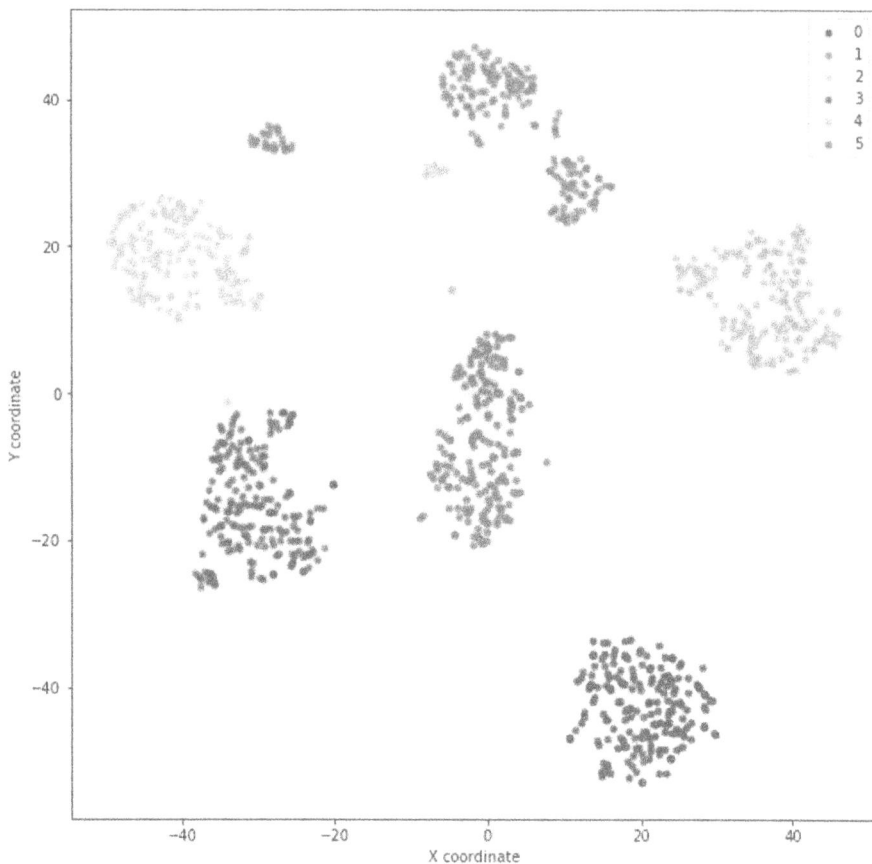

FIGURE 13.1   t-SNE on digits dataset.

```
plt.figure(figsize=(10,10))
plt.scatter(X_tsne[y==0, 0], X_tsne [y==0, 1], color='blue',
alpha=0.5,label='0', s=9, lw=1)
plt.scatter(X_tsne [y==1, 0], X_tsne [y==1, 1], color='green',
alpha=0.5,label='1',s=9, lw=1)
plt.scatter(X_tsne [y==2, 0], X_tsne [y==2, 1],
color='orange', alpha=0.5,label='2',s=9, lw=1)
plt.scatter(X_tsne [y==3, 0], X_tsne [y==3, 1],
color='purple', alpha=0.5,label='3',s=9, lw=1)
plt.scatter(X_tsne [y==4, 0], X_tsne [y==4, 1],
color='violet', alpha=0.5,label='4',s=9, lw=1)
plt.scatter(X_tsne [y==5, 0], X_tsne [y==5, 1], color='red',
alpha=0.5,label='5',s=9, lw=1)
plt.ylabel('Y coordinate')
plt.xlabel('X coordinate')
plt.legend()
plt.show()
```

Example 2

In this example, dimensionality reduction is performed by t-SNE on the MNIST dataset by taking a sample of 10000 data points to reduce the computational complexity to produce a two-dimensional mapping of the data points. This dataset is loaded from the tensorflow library and similar to the previous example, we use the implementation of the t-SNE algorithm from the sklearn library.

Import all the necessary libraries.

```
import tensorflow as tf
from tensorflow.examples.tutorials.mnist import input_data
import sklearn
from sklearn.manifold import TSNE
import matplotlib.pyplot as plt
```

Load the dataset. This dataset has 55000 data points with a dimensionality of 784. However, we are taking a sample of 10000 data points for this example.

```
mnist = input_data.read_data_sets("MNIST_data/")
X_train = mnist.train.images
y_train = mnist.train.labels
X_train = X_train [0:10000]
y_train = y_train [0:10000]
n_samples, n_features = X_train.shape
print(n_features)
print(n_samples)
```

Output:

```
784
10000
```

Use the t-SNE model from the *sklearn.manifold* module to reduce the data from 784 to 2 dimensions.

```
tsne = TSNE(n_components=2, random_state=0)
X_tsne = tsne.fit_transform(X_train)
```

Finally, plot the transformed data with each data point denoted by a color corresponding to its class label.

```
plt.figure(figsize=(10,10))
plt.scatter(X_tsne[y_train==0, 0], X_tsne[y_train==0, 1],
color='blue', alpha=0.5,label='0', s=9, lw=2)
plt.scatter(X_tsne[y_train==1, 0], X_tsne[y_train==1, 1],
color='purple', alpha=0.5,label='1',s=9, lw=2)
plt.scatter(X_tsne[y_train==2, 0], X_tsne[y_train==2, 1],
color='yellow', alpha=0.5,label='2',s=9, lw=2)
plt.scatter(X_tsne[y_train==3, 0], X_tsne[y_train==3, 1],
color='black', alpha=0.5,label='3',s=9, lw=2)
plt.scatter(X_tsne[y_train==4, 0], X_tsne[y_train==4, 1],
color='gray', alpha=0.5,label='4',s=9, lw=2)
```

```
plt.scatter(X_tsne[y_train==5, 0], X_tsne[y_train==5, 1],
color='lightgreen', alpha=0.5,label='5',s=9, lw=2)
plt.scatter(X_tsne[y_train==6, 0], X_tsne[y_train==6, 1],
color='red', alpha=0.5,label='6',s=9, lw=2)
plt.scatter(X_tsne[y_train==7, 0], X_tsne[y_train==7, 1],
color='green', alpha=0.5,label='7',s=9, lw=2)
plt.scatter(X_tsne[y_train==8, 0], X_tsne[y_train==8, 1],
color='lightblue', alpha=0.5,label='8',s=9, lw=2)
plt.scatter(X_tsne[y_train==9, 0], X_tsne[y_train==9, 1],
color='orange', alpha=0.5,label='9',s=9, lw=2)
plt.ylabel('Y coordinate')
plt.xlabel('X coordinate')
plt.legend()
plt.show()
```

The resulting low dimensional embeddings are visualized in Figure 13.2. Note that
t-SNE does a good job as the classes are clearly separated, thus revealing the natural
classes of digits in the data.

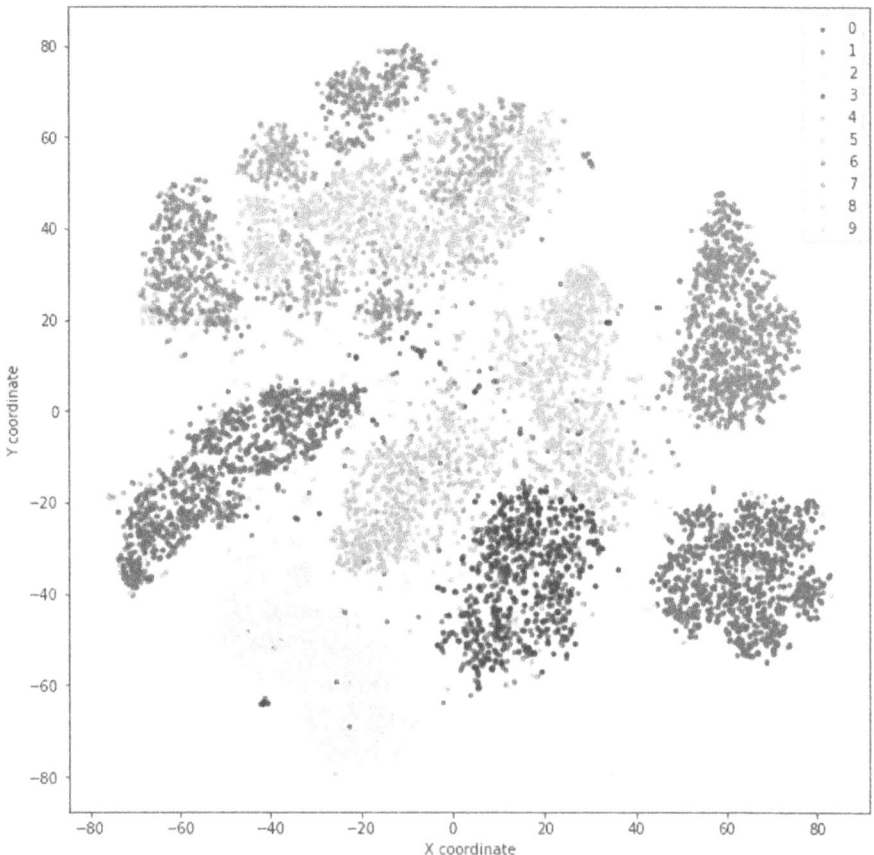

FIGURE 13.2 t-SNE on MNIST dataset.

REFERENCES

1. Maaten, L. V. D., & Hinton, G. (2008). Visualizing data using t-SNE. *Journal of machine learning research*, *9*(Nov), 2579–2605.
2. Hinton, G. E., & Roweis, S. T. (2003). Stochastic neighbor embedding. In *Advances in neural information processing systems* (pp. 857–864). Vancouver, BC, Canada.
3. Arora, S., Hu, W., & Kothari, P. K. (2018). An analysis of the t-SNE algorithm for data visualization. *arXiv preprint arXiv:1803.01768*.
4. Li, W., Cerise, J. E., Yang, Y., & Han, H. (2017). Application of t-SNE to human genetic data. *Journal of bioinformatics and computational biology*, *15*(04), 1750017.

14 Comparative Analysis of Dimensionality Reduction Techniques

14.1 INTRODUCTION

Recently, many non-linear dimensionality reduction methods have been proposed to overcome the limitations of traditional methods like Principal Component Analysis (PCA) and Multidimensional Scaling (MDS). In all previous chapters, we have discussed various unsupervised dimensionality reduction techniques, from classical methods such as PCA to state-of-the-art non-linear manifold learning techniques. This chapter will present a systematic comparison and review of each of those algorithms to better understand when and where each algorithm will be a good choice for dimensionality reduction.

Linear dimensionality reduction techniques such as PCA and classical MDS are not suitable for complex non-linear data. Real-world data being highly non-linear, these traditional linear techniques are not efficient for reducing the dimensionality of such data. However, non-linear methods overcome this limitation and are capable of handling complex non-linear data. They outperform traditional linear methods in dimensionality reduction, especially on data that lie on a highly non-linear manifold.

In this chapter, we will make a comparative study of dimensionality reduction algorithms such as PCA, MDS, Kernel PCA, Isomap, Maximum Variance Unfolding (MVU), Locally Linear Embedding (LLE), Laplacian Eigenmaps, Hessian LLE and tSNE. Apart from these, we will also briefly touch upon some more algorithms that were not covered in all previous chapters such as diffusion maps, Sammon mapping, Local Tangent Space Analysis, multilayer autoencoders, Locally Linear Coordination, and Manifold Charting.

Let X be an $n \times D$ matrix that denotes a dataset with dimensionality D and x_i be the high dimensional data points. Let the intrinsic dimensionality of the data be d; that is, the data points lie on or near a d dimensional manifold which is embedded in D-dimensional space. Hence, the aim of dimensionality reduction is to reduce the D-dimensional data X to d-dimensional data Y. Here, Y is a matrix that denotes the reduced data and y_i denotes the data points from the reduced dataset with dimensionality d.

14.1.1 DIMENSIONALITY REDUCTION TECHNIQUES

Dimensionality reduction methods can be grouped into different classes based on their working principle. One way of grouping the dimensionality reduction techniques is to classify them as convex and non-convex.

14.2 CONVEX DIMENSIONALITY REDUCTION TECHNIQUES

Convex techniques are methods that reduce the dimensionality of the data by optimizing an objective function that does not have any local optima. In convex dimensionality reduction techniques, the solution space is convex [1]. On the other hand, dimensionality reduction techniques that fall under the non-convex category are methods that optimize functions that have local optima.

Many of the dimensionality reductions that we have discussed in all previous chapters are convex techniques. Most of these methods have objective functions that can be optimized by solving a generalized eigenvalue problem.

While some of these convex techniques involve eigendecomposition of a dense matrix (full matrix), some techniques perform eigendecomposition on a sparse matrix. Hence, these convex dimensionality reduction techniques can be further classified into full spectral and sparse spectral techniques.

14.2.1 FULL SPECTRAL TECHNIQUES

Full spectral techniques perform dimensionality reduction by optimizing the objective function that can be solved by eigendecomposition of a full matrix that denotes the pairwise similarities between data points. Techniques such as PCA, MDS, Isomap, Kernel PCA, and Maximum Variance Unfolding are full spectral techniques.

PCA yields a low dimensional representation of the data by finding the subspace that retains most of the variability of the data, that is, by finding the orthogonal vectors (principal components) with maximal variance. Mathematically, PCA finds the linear mapping U which maximizes $trace(U^T cov(X)U)$. Here, $cov(X)$ is the covariance matrix of X. This linear mapping is obtained from the d principal components, which are nothing but the d principal eigenvectors of the covariance matrix. Therefore, PCA attempts to solve the following eigenvalue problem:

$$cov(X)U = \lambda U \tag{14.1}$$

This eigenvalue problem is solved for the d largest eigenvalues λ of the covariance matrix, and the principal components are nothing but the eigenvectors corresponding to the d largest eigenvalues. The low dimensional representation (Y) of the data X is obtained by mapping them to U, that is, $Y = XM$. Hence, the principal components can be found by eigendecomposition of the covariance matrix of X.

Similar to PCA, MDS attempts to find the linear mapping M which minimizes

$$\sum_{ij} \left(d_{ij}^2 - \left\| y_i - y_j \right\|^2 \right) \tag{14.2}$$

where d_{ij} denotes the entries of the Euclidean distance matrix D which represent the Euclidean distance between pairs of data points in the high dimensional space and $\left\| y_i - y_j \right\|^2$ is the squared Euclidian distance between y_i and y_j. The minimum of this cost function is nothing but the eigenvalue decomposition of a gram matrix which is a double-centered squared Euclidean distance matrix. Thus, the minimum of the cost function in (14.2) is derived by multiplying the principal eigenvectors of the gram matrix with the square root of their corresponding eigenvalues.

PCA and MDS are similar for the relation between the eigenvectors of the covariance matrix and the gram matrix respectively.

While PCA can only deal with linear data, Isomap is a manifold learning technique which can handle non-linear high dimensional data that lie on or near a manifold. Isomap attempts to preserve the pairwise geodesic distances between the data points. The geodesic distance is calculated by constructing a k-nearest neighbor graph for dataset X and the pairwise geodesic distance matrix is obtained. The lower dimensional points y_i corresponding to the high dimensional points x_i are obtained by applying multidimensional scaling on the pairwise geodesic distance matrix.

Kernel PCA is a kernel-based method which is a modified version of the traditional linear PCA in a high dimensional space which is obtained using a kernel function for constructing non-linear low dimensional mappings. While linear PCA computes the principal eigenvectors of the covariance matrix, the kernel PCA computes the principal eigenvectors of the kernel matrix.

While the mapping yielded by kernel PCA depends on the choice of the kernel function, it is uncertain how this kernel function is chosen. In order to overcome this ambiguity, Maximum Variance Unfolding (MVU) attempts to learn the kernel matrix K by constructing a k-nearest neighborhood graph on the high dimensional data X and trying to maximize the variance of the embeddings in the low dimensional feature space, that is, trying to maximize summation of the squared Euclidean distances between the data points such that the distances within the graph are preserved. The optimization problem to find the kernel is a semidefinite programming problem which is discussed in detail in Chapter 11. Thus, the low dimensional representation Y is obtained from eigenvalue decomposition of the learned kernel matrix K which is obtained by solving the semidefinite programming problem.

14.2.2 SPARSE SPECTRAL TECHNIQUES

In the previous section, we briefly discussed some of the dimensionality reduction techniques that perform eigenvalue decomposition of a full matrix to construct a low dimensional representation of high dimensional data. However, in this section, let us discuss sparse spectral techniques along with an overview of some of the techniques that come under this category of dimensionality reduction methods.

Contrary to the full spectral techniques, sparse spectral techniques obtain low dimensional representation of data by performing eigenvalue decomposition on a sparse matrix, that is, by solving a sparse generalized eigenvalue problem. These sparse spectral techniques are mainly concerned with preserving the local structure of the original high dimensional data. Some examples of dimensionality reduction techniques that are classified as sparse spectral methods are Local Linear Embedding, Laplacian Eigenmaps, Hessian Local Linear Embedding, and Local

Tangent Space Analysis. Let us discuss in brief these methods in the following subsection.

A brief overview of some sparse spectral techniques:

Local Linear Embedding (LLE) is a dimensionality reduction technique that constructs graph representation of the data, as in the case of Isomaps and MVU. However, unlike Isomap, LLE focuses merely on preserving the local structure of the data. This is done by constructing a neighborhood graph and then finding the weights for reconstructing the data in that neighborhood and finally computing the coordinates best constructed by those weights in lower dimension such that the local properties of the data are retained as well as possible. The optimization problem to be solved for obtaining the low dimensional representation is as follows:

$$\min_{y} \sum_{i}^{n} \left| y_i - \sum_{j=1}^{k} w_{ij} y_j \right|^2 \tag{14.3}$$

The low dimensional mapping that minimizes this cost function can be found by calculating the $d+1$ eigenvectors of $(I-W)(I-W)^T$, that is, the eigenvectors corresponding to the lowest d nonzero eigenvalues. Here, W is an $n \times n$ sparse matrix representing the weights.

Similar to LLE, Laplacian Eigenmap attempts to retain the local properties of the manifold on which the data lie while obtaining the low dimensional representation. Minimizing the cost function is defined as an eigenvalue problem using spectral graph theory. Thus, by solving the generalized eigenvalue problem $Ly = \lambda Dy$ for the d smallest nonzero eigenvalues where D is the degree matrix and L is the graph Laplacian matrix, the low dimensional representation Y can be obtained.

Hessian Local Linear Embedding is a variant of the standard LLE. This method minimizes the curviness of the manifold while mapping it to the lower dimensional space in such a way that the low dimensional representation of the data is locally isometric. The representation of the data in low dimensional space is obtained by an eigenanalysis of a matrix H which represents the manifold's curviness around the data points. The manifold's curviness is measured by the local Hessian at each point which is represented in the local tangent space at the data point. The low dimensional representation is obtained by finding the eigenvectors corresponding to the d smallest non-zero eigenvalues of H. Hessian LLE is similar to Laplacian Eigenmaps, for it just replaces the manifold Laplacian by manifold Hessian.

Like Hessian LLE, Local Tangent Space Analysis (LTSA) defines the local properties of high dimensional data using the local tangent space of every data point [2]. There is a linear mapping from a data point in the high dimensional space to its local tangent space and likewise, there exists a linear mapping from the corresponding data point in the low dimensional space to that local tangent space if local linearity of the manifold is assumed. LSTA aligns these mappings such that the local tangent space of the manifold is constructed from the low dimensional representation.

14.3 NON-CONVEX TECHNIQUES FOR DIMENSIONALITY REDUCTION

In the previous section, we discussed convex dimensionality reduction techniques that attempt to optimize convex objective function by performing eigenvalue decomposition to construct low dimensional representations. However, in this section, we will briefly discuss non-convex techniques for dimensionality reduction. Some of the non-convex dimensionality reduction techniques are Sammon mapping, Multilayer Autoencoders, Locally Linear Coordination, and Manifold Charting.

Sammon mapping is a variant of MDS which focuses mainly on preserving the small pairwise distances between the data points to retain the geometry of the data. Sammon mapping modifies the MDS cost function in (13.2) in such a way that the influence of each pair (i, j) is weighted to the cost function by the inverse of their corresponding pairwise Euclidean distance d_{ij}. The Sammon cost function is as follows:

$$\varnothing(Y) = \frac{1}{\sum_{ij} d_{ij}} \sum_{i \neq j} \frac{\left(d_{ij} - \|y_i - y_j\|\right)^2}{d_{ij}} \tag{14.4}$$

where d_{ij} denotes the Euclidean distance between the pairs of high dimensional data points. The constant $\frac{1}{\sum_{ij} d_{ij}}$ is added to it for simplifying the gradient of the cost function. Therefore, the modified cost function allocates approximately equal weight to preserving the distances between pairs of data points and hence preserves the local structure of the data more efficiently than multidimensional scaling. Sammon mapping is most commonly used as a technique for data visualization [3]. However, Sammon mapping is also applied on gene data and geospatial data [4].

Next, let us look into Multilayer Autoencoders. Multilayer Autoencoders are nothing but feed forward neural networks with an odd number of hidden layers [5, 6] with weights shared between the top and bottom hidden layers. While the input and output layers of the neural network have D nodes, the middle hidden layer has d nodes. This feed-forward neural network architecture is trained in such a way that the mean squared error between the input and the output is minimized. By training the network on the high dimensional data points, the d-dimensional representation of the data which retains the original geometry of the data is obtained from the middle hidden layers. Using the high dimensional data point x_i as the input to the neural network, its corresponding low dimensional embedding y_i can be obtained from the node values in the middle hidden layer. A commonly used activation function is the sigmoid activation function that facilitates non-linear mapping of the high dimensional data points to their lower dimensional counterparts. However, linear activation function is usually used for the middle hidden layers. Since Multilayer Autoencoders have many connections, the convergence of backpropagation methods are slow and can possibly get stuck in local minima. However, a three-stage training procedure in which the network learns in three main stages can overcome the limitation of being susceptible to local minima [7].

Locally Linear Coordination [8] is a technique that computes many locally linear models and further makes a global alignment of these locally linear models. Firstly, the Expectation-maximization algorithm is employed to compute a mixture of locally linear models on the data. Next, the local linear models are aligned to obtain the low dimensional representation of the data using a variant of the LLE algorithm. This technique attempts to solve a generalized eigenvalue problem similar to LLE in order to obtain the low dimensional data representation.

Similar to Locally Linear Coordination, the Manifold Charting technique obtains the low dimensional mappings by aligning local linear models. But unlike LLC which minimizes the cost function of LLE, Manifold Charting does not attempt to minimize a cost function of any other dimensionality reduction method. However, it minimizes a convex cost function which quantifies the degree of disagreement between the linear models by solving a generalized eigenproblem.

14.4 COMPARISON OF DIMENSIONALITY REDUCTION TECHNIQUES

In the previous sections, we discussed the categorization of dimensionality reduction techniques based on their working principle and provided a brief overview of some of the techniques that come under each category. Now, let us discuss the relations among these dimensionality reduction techniques and compare and contrast their general properties.

Many of the dimensionality reduction techniques that we have seen earlier are highly comparable and interrelated and sometimes even equivalent. Firstly, traditional linear PCA is in many ways equivalent to classical scaling and kernel PCA using a linear kernel. They are identical with regard to the similarity between the eigenvectors of the covariance matrix and the double-centered squared Euclidean distance matrix (as in the case of MDS) which is again equivalent to the Gram matrix. Moreover, Multilayer Autoencoders that use only linear activation functions for their network layers are very comparable to linear PCA [9].

Likewise, carrying out multidimensional scaling on a pairwise geodesic distance matrix is similar to the technique of Isomap. When Isomap is used for dimensionality reduction with the number of nearest neighbors set to $n - 1$, it becomes equivalent to linear PCA, classical scaling, and kernel PCA using a linear kernel. While diffusion maps are identical to multidimensional scaling, they retain "diffusion" distances which are different from the type of pairwise distances preserved by MDS.

Spectral techniques such as the kernel PCA, LLE, Isomap, and Laplacian Eigenmap can be considered special cases of a more general problem of learning eigenfunctions [10]. Therefore, Laplacian Eigenmaps, Isomap, and LLE are taken to be special cases of kernel PCA with a particular kernel function K. As stated earlier, Hessian LLE and Laplacian Eigenmap are also closely related with the only difference being the differential operator defined on the manifold (manifold Hessian and manifold Laplacian respectively).

On the other hand, MVU can also be considered as a special case of kernel PCA in which the kernel function is the solution of the semidefinite programming problem.

Then, Isomap can be considered to be a method that can find an approximate solution to the MVU problem [11]. These relations between the techniques make them interrelated and similar in terms of their underlying principle.

By studying the properties of these techniques, we can say that most of these dimensionality reduction techniques are non-parametric, which implies that these methods do not define a direct mapping from the high dimensional space to the lower dimensional space as well as from the low dimensional space to the higher dimensional space. Because of this, we cannot generalize to new test data without carrying out the dimensionality reduction on the data again, and at the same time it is not possible to find the error between the true data and the reconstructed data; that is, we cannot estimate how much information from the high dimensional space has been retained in its low dimensional embedding.

Reviewing these techniques, it can be noted that most of the non-linear dimensionality reduction methods involve objective functions that have free parameters which have to be optimized; that is, they have parameters that have a direct influence on the cost function to be optimized. While free parameters make the techniques more flexible, they can be disadvantageous as they need to be tuned for optimal performance of the technique.

Now, let us take a closer look at the memory and computational complexities of these dimensionality reduction techniques. This aspect of a technique is very crucial to determine its efficiency and practical applicability. Supposing the memory and computational requirements required for performing dimensionality reduction on a dataset is very large; it would become infeasible for practical application. The characteristics to be considered to determine the computational complexity of a dimensionality reduction technique are:

   a. The nature of the data such as its original dimensionality D, number of data points n, etc.
   b. The parameters of the dimensionality reduction method such as the number of nearest neighbors in the neighborhood graph, the target dimensionality to be achieved, etc.

The most computationally expensive component of the PCA algorithms is the eigenanalysis of the covariance matrix with a computational complexity of $O(D^3)$ and memory complexity of $O(D^2)$. However, when the number of data points is less than the original dimensionality, that is, if $n < D$, then both the computational and memory complexity of linear PCA reduces to $O(n^3)$ and $O(n^2)$ respectively. On the other hand, for techniques including MDS, Kernel PCA and Isomap, the most computationally expensive component is the eigenanalysis of $n \times n$ matrix which has a computational complexity of $O(n^3)$. The memory complexity is $O(n^2)$ since these full spectral techniques involve a full $n \times n$ matrix.

Since sparse spectral techniques perform eigenanalysis of a sparse $n \times n$ matrix, the computational complexity is less. Hence, the computational and memory complexity of eigenanalysis of the sparse matrix is $O(pn^2)$ where p is the ratio of the nonzero entries to the total entries in the sparse matrix.

The comparative analysis of the dimensionality reduction techniques infers the following about these techniques:

1. Most of the non-linear dimensionality reduction techniques that we have disused are non-parametric. That is, they do not define a direct mapping between the high dimensional and low dimensional space.
2. Every non-linear technique involves one or many free parameters which need to be optimized.
3. When the original dimensionality D of the data is smaller than the number of number of data points n, that is, if $D < n$, then non-linear techniques are computationally less efficient as compared to linear PCA.
4. Many of the non-linear dimensionality reduction techniques have memory complexity which is the square or cube of the number of data points n.

From the above stated observations, it can be inferred that non-linear dimensionality reduction techniques are computationally demanding when compared to PCA. However, many methods have been proposed as an attempt to make these algorithms more efficient in terms of their computational and memory requirements.

14.5   COMPARISON OF MANIFOLD LEARNING METHODS WITH EXAMPLE

In the previous sections, we briefly discussed and analyzed the working principle, general properties, and relations between various dimensionality reduction techniques. In this section, we will be comparing the different manifold learning techniques discussed in this chapter with the help of an example. In this example, let us obtain and visualize the low dimensional embeddings of a dataset using different manifold learning techniques.

Firstly, we will illustrate the dimensionality reduction of a spherical dataset with various manifold learning methods. Here we can observe the use of dimensionality reduction to understand manifold learning methods. In this severed sphere dataset, the poles of the sphere and a thin slice along its side are cut from the sphere in such a way that the manifold learning techniques can spread the sphere open while projecting onto the two-dimensional space. In this example, we will be comparing manifold learning techniques including LLE, Hessian LLE, Modified LLE, Isomap, MDS, Laplacian Eigenmaps, t-SNE, and Local Tangent Space Analysis.

To create this dataset let us use the scikit-learn library. Likewise, we will use the *sklearn,manifold* module from the scikit-learn library which provides the implementation of the manifold learning techniques.

First let us import all the necessary libraries.

```
import numpy as np
from sklearn import datasets
from sklearn import manifold
```

```
import matplotlib.pyplot as plt
from matplotlib.ticker import NullFormatter
from sklearn.utils import check_random_state
from mpl_toolkits.mplot3d import Axes3D
```

Next, let us create the sphere dataset with 2000 datapoints.

```
n_neighbors = 10
n_points = 2000
random_state = check_random_state(0)
p = random_state.rand(n_points) * (2 * np.pi - 0.55)
t = random_state.rand(n_points) * np.pi
```

Once the sphere dataset is created, we have to sever the sphere.

```
indices = ((t < (np.pi - (np.pi / 8))) & (t > ((np.pi / 8))))
colors = p[indices]
x, y, z = np.sin(t[indices]) * np.cos(p[indices]),
np.sin(t[indices]) * np.sin(p[indices]), np.cos(t[indices])
```

After slicing the poles and a small strip from the sides of the sphere, plot the resulting severed dataset.

```
fig = plt.figure(figsize=(45, 20))
ax = fig.add_subplot(251, projection='3d')
ax.scatter(x, y, z, c=p[indices], cmap=plt.cm.jet)
ax.view_init(40, -10)
sphere_data = np.array([x, y, z]).T
```

Figure 14.1 is the scatter plot of the severed sphere data with 2000 data points.

Now that we have the dataset, let us reduce the dimensionality of the data to two using manifold learning methods from sklearn. First, we will perform dimensionality reduction with standard LLE using *sklearn,manifold. LocallyLinearEmbedding* with number of neighbors as 10, target projection space as 2, and method as "standard." After fitting and transforming the data, plot the results using matplotlib.

```
X_lle= manifold.LocallyLinearEmbedding(n_neighbors=10, n_
components=2,method='standard').fit_transform(sphere_data).T
fig = plt.figure(figsize=(10, 10))
ax = fig.add_subplot(1,1,1)
plt.scatter(X_lle[0], X_lle[1], c=colors, cmap=plt.cm.jet)
plt.title("MDS")
ax.xaxis.set_major_formatter(NullFormatter())
ax.yaxis.set_major_formatter(NullFormatter())
plt.axis('tight')
plt.ylabel('Y coordinate')
plt.xlabel('X coordinate')
plt.show()
```

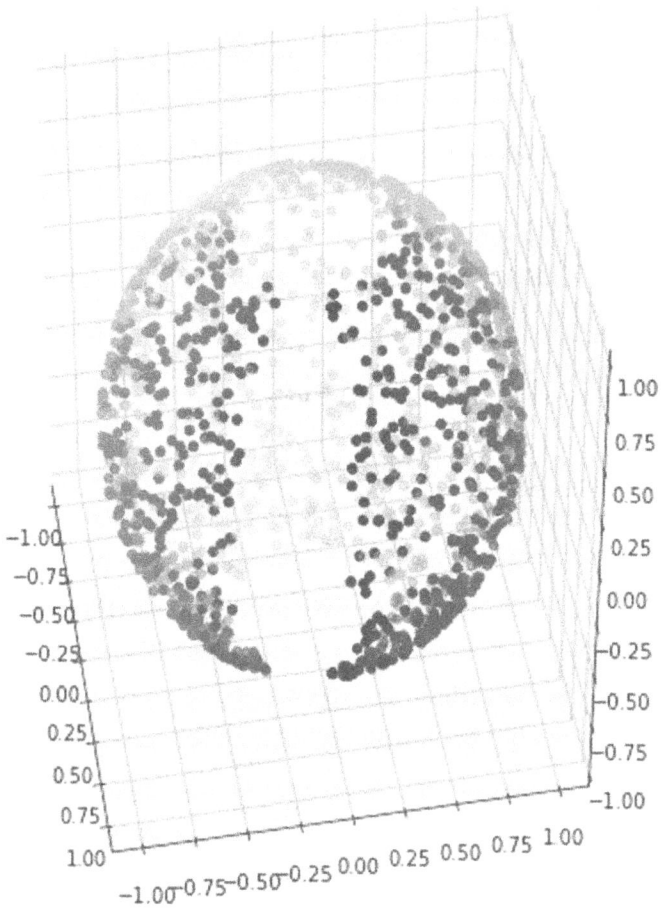

FIGURE 14.1   Severed sphere data (n = 2000).

Similarly, for modified LLE, use *sklearn,manifold.LocallyLinearEmbedding* with number of neighbors as 10, target projection space as 2, and method as "modified."

```
X_mod_lle= manifold.LocallyLinearEmbedding(n_neighbors=10,
n_components=2,method='modified').fit_transform(sphere_data).T
```

For Hessian LLE, use *sklearn,manifold.LocallyLinearEmbedding* with number of neighbors as 10, target projection space as 2, and method as "hessian."

```
X_hess_lle= manifold.LocallyLinearEmbedding(n_neighbors=10,
n_components=2,method='hessian').fit_transform(sphere_data).T
```

For LTSA, use *sklearn,manifold.LocallyLinearEmbedding* with number of neighbors as 10, target projection space as 2, and method as "ltsa."

```
X_lsta= manifold.LocallyLinearEmbedding(n_neighbors=10, n_
components=2,method='ltsa').fit_transform(sphere_data).T
```

For Isomap, use *sklearn,manifold.Isomap* with number of neighbors as 10 and target projection space as 2.

```
X_iso= manifold.LocallyLinearEmbedding(n_neighbors=10, n_
components=2).fit_transform(sphere_data).T
```

For MDS, use *sklearn,manifold.MDS* with target projection space as 2.

```
X_mds= manifold.MDS(n_components=2).
fit_transform(sphere_data).T
```

For Laplacian Eigenmaps, use *sklearn,manifold.SpectralEmbedding* with number of neighbors as 10 and target projection space as 2.

```
X_le= manifold.SpectralEmbedding(n_neighbors=10, n_
components=2).fit_transform(sphere_data).T
```

For t-SNE, use *sklearn,manifold.TSNE* with target projection space as 2.

```
X_tsne= manifold.TSNE(n_components=2,init='pca',random_
state=0).fit_transform(sphere_data).T
```

Now, plot the results of the dimensionality reduction by the manifold learning algorithms using matplotlib. The visualizations of the resultant low dimensional representation of the data by each manifold learning algorithm are illustrated in Figure 14.2.

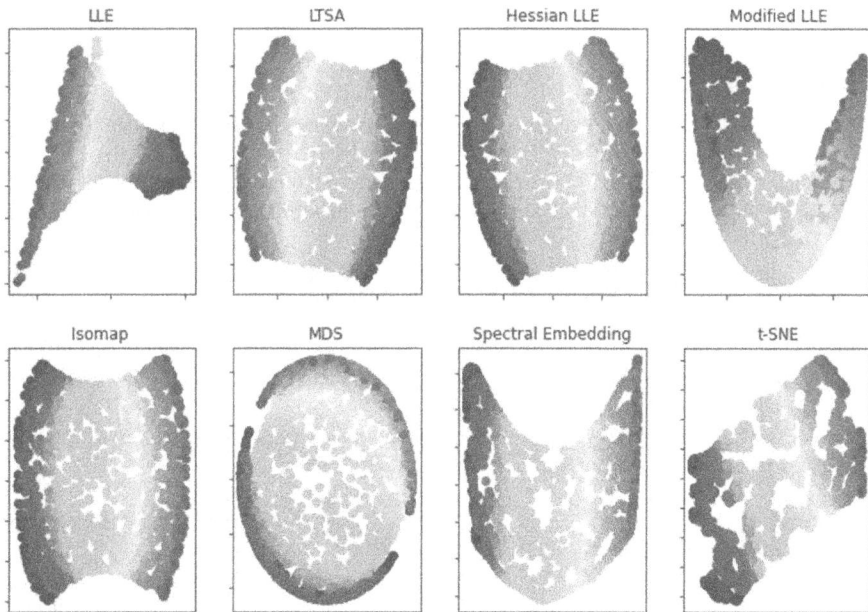

FIGURE 14.2 Manifold learning algorithms on severed sphere dataset.

14.6   DISCUSSION

In this chapter, we have reviewed and compared various techniques for dimensionality reduction. Full spectral techniques that use neighborhood graphs such as Isomaps and MVU are susceptible to the limitations such as overfitting, the curse of dimensionality, and the presence of outliers. Apart from that, we have already discussed the problem of short-circuiting in Isomaps and MVU. On the other hand, kernel-based techniques are not subject to any of the drawbacks of neighborhood graph-based dimensionality reduction techniques. However, methods like kernel PCA are not very efficient in modeling a complex non-linear manifold.

Furthermore, all of the sparse spectral techniques suffer from the curse of dimensionality of the embedded manifold since the number of data points needed to describe a manifold accurately increases exponentially with the manifold's intrinsic dimensionality. For datasets with low intrinsic dimensionality such as the S-curve data, the severed sphere dataset, and the Swiss roll data this limitation is not observed. However, since most of the real-world data have very high intrinsic dimensionality, this becomes a concern. Another point to be noted is that a manifold's local properties need not essentially follow its global structure when noise is present around the manifold. Thus, sparse spectral techniques are vulnerable to overfitting on the manifold. In addition to this, sparse spectral techniques are not capable of modeling non-smooth manifolds as they assume local linearity; that is, they assume that the underlying manifold is smooth without any irregularities. While we have discussed and compared various important dimensionality reduction techniques in this chapter, it is not exhaustive. There are many other methods that can be used for reducing the dimensionality of a given dataset.

REFERENCES

1. Boyd, S. P., & Vandenberghe, L. (2004). *Convex optimization*. Cambridge University Press, New York, NY.
2. Zhang, Z., & Zha, H. (2004). Principal manifolds and nonlinear dimensionality reduction via local tangent space alignment. *SIAM journal of scientific computing*, 26(1), 313–338.
3. Martin-Merino, M., & Munoz, A. (2004). A new Sammon algorithm for sparse data visualization. In *Proceedings of the 17th international conference on pattern recognition* (pages 477–481). Cambridge, UK.
4. Takatsuka, M. (2001). An application of the self-organizing map and interactive 3-D visualization to geospatial data. In *Proceedings of the 6th international conference on GeoComputation*. Brisbane, Australia.
5. DeMers, D., & Cottrell, G. (1993). Non-linear dimensionality reduction. In *Advances in neural information processing systems* (volume 5, pages 580–587). Morgan Kaufmann, San Mateo, CA.
6. Hinton, G. E., & Salakhutdinov, R. R. (2006). Reducing the dimensionality of data with neural networks. *Science*, 313(5786), 504–507.
7. Larochelle, H., Bengio, Y., Louradour, J., & Lamblin, P. (2009). Exploring strategies for training deep neural networks. *Journal of machine learning research*, 10(Jan), 1–40.

8. Teh, Y. W., & Roweis, S. T. (2002). Automatic alignment of hidden representations. In *Advances in neural information processing systems*, volume 15, pages 841–848, The MIT Press, Cambridge, MA.

9. Kung, S. Y., Diamantaras, K. I., & Taur, J. S. (1994). Adaptive principal component EXtraction (APEX) and applications. *IEEE transactions on signal processing*, 42(5), 1202–1217.

10. Bengio, Y., Delalleau, O., Le Roux, N., Paiement, J.-F., Vincent, P., & Ouimet, M. (2004). Learning eigenfunctions links spectral embedding and kernel PCA. *Neural computation*, 16(10), 2197–2219.

11. Xiao, L., Sun, J., & Boyd, S. (2006). A duality view of spectral methods for dimensionality reduction. In *Proceedings of the 23rd international conference on machine learning* (pages 1041–1048). Pittsburgh, PA.

Glossary of Words and Concepts

Backpropagation: Backpropagation refers to the backward propagation of errors while training neural networks using gradient descent. In a backpropagation algorithm the error of the cost function is propagated backwards through the neural network layers in order to adjust the weights such that the error is minimized.

Bivariate correlation between sets of variables: Simple bivariate correlation is a statistical technique that is used to determine the existence of relationships between two different variables (i.e., X and Y). It shows how much X will change when there is a change in Y.

Categorical variable: A categorical variable is a variable that can take on one of a limited, and usually fixed, number of possible values, assigning each individual or other unit of observation to a particular group or nominal category on the basis of some qualitative property.

Constrained Optimization Problems: Constrained optimization problems are optimization problems where the objective function is minimized or maximized subject to constraints.

Convex function: A convex function is a continuous function whose value at the midpoint of all the intervals in the function's domain is lesser than the mean of its values at the ends of the interval. In other words, a function is a convex function if a line segment joining any two points on the function's graph lies above the graph between those two points.

Convex optimization problem: A convex optimization problem is an optimization problem where all the constraints are convex functions and the objective function of the optimization problem is a convex function which is minimized.

Correlated features: It often happens that two features (correlated features) that were meant to measure different characteristics are influenced by some common mechanism and tend to vary together. For example, the perimeter and the maximum width of a figure will both vary with scale; larger figures will have both larger perimeters and larger maximum widths.

Correlation matrix: A correlation matrix is simply a table which displays the correlation coefficients for different variables. The matrix depicts the correlation between all the possible pairs of values in a table.

Cost function: A cost function or a loss function is nothing but an objective function whose value is to be minimized in an optimization problem.

Covariance matrix: A covariance matrix (also known as auto-covariance matrix, dispersion matrix, variance matrix, or variance-covariance matrix) is a square matrix giving the covariance between each pair of elements of a given random vector.

Covariance/Covariance normalized by variance (correlation coefficient): In probability theory and statistics, covariance is a measure of the joint variability of two random variables. If the greater values of one variable mainly correspond with the greater values of the other variable, and the same holds for the lesser values (that is, the variables tend to show similar behavior), the covariance is positive. In the opposite case, when the greater values of one variable mainly correspond to the lesser values of the other (that is, the variables tend to show opposite behavior), the covariance is negative. The sign of the covariance therefore shows the tendency in the linear relationship between the variables. The magnitude of the covariance is not easy to interpret because it is not normalized and hence depends on the magnitudes of the variables. The normalized version of the covariance, the correlation coefficient, however, shows by its magnitude the strength of the linear relation.

Curse of dimensionality: Curse of dimensionality basically refers to the difficulties that arise while working with high dimensional data. It implies that the error increases with the increase in the dimensionality of the data (with the increase in the number of features in the data).

Data standardization: Standardizing a dataset involves rescaling the distribution of values so that the mean of observed values is 0 and the standard deviation is 1. It is sometimes referred to as "whitening."

Dense matrix: A matrix where most of its entries are non-zero is called a dense matrix.

Diagonal matrix: Diagonal matrix is a matrix whose entries except the main diagonal entries are all zero.

Dijkstra's algorithm: Dijkstra's algorithm is a step-by-step process we can use to find the shortest path between two vertices in a weighted graph. This algorithm enables us to find shortest distances and minimum costs.

Dissimilarity matrix: The dissimilarity matrix (also called distance matrix) describes a pairwise distinction between M objects. It is a square symmetrical M × M matrix with the (i, j)th element equal to the value of a chosen measure of distinction between the (i)th and the (j)th object. The diagonal elements are either not considered or are usually equal to zero – i.e. the distinction between an object and itself is postulated as zero.

Distance matrix: In mathematics, computer science, and especially graph theory, a distance matrix is a square matrix (two-dimensional array) containing the distances, taken pairwise, between the elements of a set.

Eigendecomposition: Eigendecomposition is a type of matrix decomposition which involves decomposing or factorizing a square matrix into a canonical form by which it is represented in terms of its eigenvalues and eigenvectors. In simple terms, eigendecomposition refers to the decomposition of a square matrix into a set of eigenvalues and eigenvectors. Eigendecomposition of a matrix is important as it is used for finding the maximum and minimum of functions involving the matrix, such as in Principal Component Analysis (PCA). Eigendecomposition is also referred as eigenvalue decomposition or spectral decomposition.

Eigenvalue problem: *Standard eigenvalue problem:* The eigenvalue problem of a d × d symmetric matrix denoted by $A \in R^{d \times d}$ is defined as $Av = \lambda v$ where the diagonal elements of λ are the eigenvalues and the columns of v are the eigenvectors. For symmetric matrix A, its eigenvectors are orthogonal/orthonormal. However, for an eigenvalue problem the matrix A can be a non-symmetric matrix as well. In that case, the eigenvectors are not orthogonal/orthonormal.

Generalized eigenvalue problem: The generalized eigenvalue problem of two d × d symmetric matrices $A \in R^{d \times d}$ and $B \in R^{d \times d}$ denoted by (A, B) is defined as $Av = \lambda Bv$ where the diagonal elements of λ are the eigenvalues of the pair (A, B) and the columns of v are the eigenvectors. This is called a generalized eigenvalue problem. Note that the standard eigenvalue problem is a special case of the generalized eigenvalue problem where $B = I$.

Expectation-maximization: The expectation-maximization algorithm finds the maximum-likelihood estimates for model parameters in the presence of latent variables. It is an iterative algorithm that first estimates the values for the latent variable (E-step) and then optimizes the model (M-step). The E and M steps are carried out iteratively until convergence.

Exploratory data analysis: Exploratory data analysis refers to the process of analyzing data using graphical representations and data visualization techniques to know the characteristics of the data such as identifying patterns, anomalies, and relationships among variables.

Feature extraction method: Usually, the techniques to obtain dimensionality reduction fall into two classes: Feature extraction – Creating new independent variables, where each independent variable is a combination of the given old independent variables; Feature elimination – Dropping features to reduce the feature space. And due to this, information is lost due to eliminated features.

Feature space: A feature vector is an n-dimensional vector of numerical features that represent some object. The vector space associated with these vectors is often called the feature space.

Floyd—Warshall algorithm: In computer science, the Floyd–Warshall algorithm (also known as Floyd's algorithm, the Roy–Warshall algorithm, the Roy–Floyd algorithm, or the WFI algorithm) is an algorithm for finding shortest paths in a directed weighted graph with positive or negative edge weights (but with no negative cycles). A single execution of the algorithm will find the lengths (summed weights) of shortest paths between all pairs of vertices. Although it does not return details of the paths themselves, it is possible to reconstruct the paths with simple modifications to the algorithm.

Gradient descent: Gradient descent is an optimization algorithm used to find the local minima of a differentiable function. It is widely used for training machine learning models where the algorithm iteratively tweaks the parameters in order to minimize the given function to its local minimum.

Graph Laplacians: For a given graph G, let A and B be the adjacency matrix and incidence matrix of G respectively and D be the degree matrix. Then, $BB^T = D - A$. The matrix, $L = D - A$ is symmetric, positive semidefinite and is called the Graph Laplacian of graph G.

Graph partitioning problem: Graph partitioning refers to the partitioning of a graph's set of nodes into mutually exclusive groups. The objective of the graph partitioning problem is to partition graph G into k components in such a way that the sum of the node weights in each component is distributed evenly and the sum of edge weights of the edges connecting the partitions is at minimum.

Grayscale: Grayscale refers to the range of monochromatic shades from black to white. Grayscale images contain only black, white, and gray shades and no other color. In this type of image, each pixel is represented by 8 bits and the combination of these 8 binary digits represents the pixel value. Hence, there are 256 grayscale levels with the pixel value ranging from 0-255 where 0 represents the darkest black and 255 represents the brightest white.

Hyperparameter: In machine learning, a hyperparameter is a parameter whose value is used to control the learning process. By contrast, the values of other parameters (typically node weights) are derived via training.

Inner product: Inner product is an operation on two vectors that produces a scalar. Inner product is a generalization of dot product and it is the multiplication of two vectors together in a vector space which results in a scalar. It is denoted by angle brackets $\langle a | b \rangle$.

Ill-defined problems/Well-defined problems: The ill-defined problems are those that do not have clear goals, solution paths, or expected solutions. The well-defined problems have specific goals, clearly defined solution paths, and clear expected solutions.

Lagrange's multipliers: The method of Lagrange multipliers is a strategy for finding the local maxima and minima of a function subject to the condition that one or more equations have to be satisfied exactly by the chosen values of the variables. The method can be summarized as follows: in order to find the maximum or minimum of a function f(x) subjected to the equality constraint g(x) = 0, form the Lagrangian function. L(x, λ) = f(x) − λ g(x) and find the stationary points of L considered as a function of x and the Lagrange multiplier λ. The great advantage of this method is that it allows the optimization to be solved without explicit parameterization in terms of the constraints. As a result, the method of Lagrange multipliers is widely used to solve challenging constrained optimization problems.

Least squares problem: The ordinary least squares is used for estimating the unknown parameters in a linear regression model by minimizing the sum of the squares of the differences between the observed values of the dependent variable and the predicted values of the independent variable.

Manifold: A manifold is an n-dimensional space where at each point there exists a neighborhood that resembles n-dimensional Euclidean space. Intuitively, we can imagine a manifold as a space which, upon magnification, is comparable to a zero curvature Euclidean space (similar to an infinite plane). Hence, a manifold is an abstract mathematical space whose local geometry at every point is analogous to Euclidean space, irrespective of its global structure.

Neighborhood: In graph theory, the neighborhood of a vertex v in graph $G = (V, E)$ is nothing but the subgraph of G that is composed of all the adjacent vertices of v and all the edges connecting those vertices to v.

Objective function: An objective function is a real-valued function whose value needs to be maximized or minimized over a defined set of input values in order to solve an optimization problem. They define the objective of the optimization problem.

Optimization problem: In simple terms, an optimization problem is the problem of finding the optimal solution from all the possible solutions.

Orthogonal vectors: We say that two vectors are orthogonal if they are perpendicular to each other; i.e., the angle between them is 90°.

Orthonormal vector: Two vectors are said to be orthogonal if they are at right angles to each other (their dot product is zero). A set of vectors is said to be orthonormal if they are all normal, and each pair of vectors in the set is orthogonal. Orthonormal vectors are usually used as a basis on a vector space.

Outliers: Outliers are data points in a dataset that are significantly different from other observations. In simple terms, they are unusual values in the dataset.

Overfitting: Overfitting refers to a model that models the training data too well. Overfitting happens when a model learns the detail and noise in the training data to the extent that it negatively impacts the performance of the model on new data.

Positive semi-definite matrix: A positive semi-definite matrix is a symmetric matrix whose eigenvalues are real and non-negative.

Projection/projection matrix: Projection matrix P is a d $\times$ d square matrix which gives the vector space projection from R^d to a subspace. It maps the vector of dependent variable values to the vector of predicted values. A projection matrix is symmetric and idempotent. That is, $P' = P$ and $P^2 = P$.

Regularization: Regularization refers to the technique of tuning a function by adding an additional penalty term to the error function. It is the modification done to a learning algorithm so that the model generalizes better. It is usually done to avoid overfitting or solve an ill-posed problem. The penalty term added to the error function to impose a cost on the optimization function for overfitting is referred to as the regularization term.

Semi-definite programming: Semi-definite programming is a subclass of convex optimization where the objective function is linear with the constraints being linear equalities or inequalities. The linear objective function is optimized over the intersection of the cone of positive semi-definite matrices with a spectrahedron (an affine space).

Singular value decomposition: SVD is a matrix decomposition technique that factorizes a matrix into component matrices. It generalizes the eigendecomposition of a square matrix to an m $\times$ n matrix. The SVD of an m $\times$ n matrix A is the factorization $A = U\Sigma V^*$ where U and V are m $\times$ m and n $\times$ n unitary matrices, respectively, and Σ is an m $\times$ n diagonal matrix.

Sparse matrix: A matrix with most of its entries being zero is called a sparse matrix.

Spectral graph theory: Spectral graph theory is the study of the properties of graphs in connection with matrices such as the adjacency matrix or Laplacian matrix along with their eigenvalues, eigenvectors, and characteristic polynomial.

Standardized input matrix: Standardizing a dataset involves rescaling the distribution of values so that the mean of observed values is 0 and the standard deviation is 1. It is sometimes referred to as "whitening."

Subspace: A subspace, also known as a vector subspace is a vector space which is a part of another vector space. That is, a subspace is nothing but a vector space which is a subset of a larger vector space.

Support-Vector Machines: SVMs, also called support-vector networks, are supervised learning models with associated learning algorithms that analyze data for classification and regression analysis.

t-distribution: t-distribution, which is also called as the student's t-distribution, is a probability distribution similar to the bell-shaped normal distribution but with heavier tails. This implies that t-distribution has a greater probability for extreme values as compared to normal distribution, as a result of which it has fatter tails. It is also known as heavy-tailed distribution.

Type I error: In statistical hypothesis testing, a type I error is the rejection of a true null hypothesis (also known as a "false positive" finding or conclusion; example: "an innocent person is convicted"), while a type II error is the non-rejection of a false null hypothesis (also known as a "false negative" finding or conclusion; example: "a guilty person is not convicted").

Variability: Variability (also called spread or dispersion) refers to how spread out a set of data is. Variability is the difference being exhibited by data points within a data set, as related to each other or as related to the mean.

Well-defined optimization problem-Simplified optimization problem: The ill-defined problems are those that do not have clear goals, solution paths, or expected solutions. The well-defined problems have specific goals, clearly defined solution paths, and clear expected solutions.

Index

For Product Safety Concerns and Information please contact our EU
representative GPSR@taylorandfrancis.com
Taylor & Francis Verlag GmbH, Kaufingerstraße 24, 80331 München, Germany